Home Wiring

Home Wiring

GUILD OF
MASTER CRAFTSMAN
PUBLICATIONS

Phil Thane

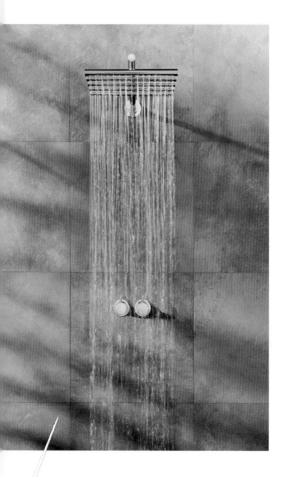

First published 2009 by
Guild of Master Craftsman Publications Ltd
Castle Place, 166 High Street,
Lewes, East Sussex BN7 1XU

Text © Phil Thane 2009
© in the Work GMC Publications 2009

ISBN 978-1-86108-627-3

A catalogue record for this book is available from the British Library.

Associate Publisher Jonathan Bailey
Production Manager Jim Bulley
Managing Editor Gerrie Purcell
Editor Mark Bentley
Managing Art Editor Gilda Pacitti
Designer Terry Jeavons
Photographer Anthony Bailey (except where stated on page 126)
Illustrator John Woodcock (except pages 22 and 99 by Gilda Pacitti)
Consultant Dr David Kennaird

Set in Clarendon and Meta Plus
Colour origination by GMC Reprographics
Printed and bound in Thailand by Kyodo Nation Printing

Contents

3 Indoors and Out

pages 88–109

Troubleshooting and the Law

pages 110–119

4

Introduction

If you work to the same standards as a professional, which you can if you follow the instructions carefully, you should have no problem.

The law in Scotland is somewhat different, but again providing you work to professional standards you should have no problem. Northern Ireland has no such regulation, but that's no reason to kill yourself.

Electric lighting and electrical appliances are central to our lives, and the ability to modify, extend or replace the circuits in your home should be a part of any keen DIYers repertoire. Having the right lighting can make a real difference to the feel of a room, and few houses have enough sockets in the right places for all the gadgets. Changing lighting systems and adding sockets isn't hard, and the range of fittings available these days to personalize your home is vast, so enjoy your electricals.

DIY electrical work has always been popular, it doesn't involve heavy lifting like many building jobs nor rely on a high level of skill like plastering. Anyone with a screwdriver and a pair of pliers can fiddle with wiring, and many have – sometimes fatally.

Because of the potential dangers, recent changes to the building regulations in England and Wales have imposed restrictions on how electrical work is done, and how it is checked and approved. But, contrary to many people's belief, there is no ban on DIY domestic electrical work, just a ban on sloppy and dangerous work.

WARNING

Remember, electricity can kill. Before you undertake ANY electrical work, please read pages 117–119 (which outlines the regulations and inspection procedures).

If you are unsure of how these apply to your project, contact your local authority.

For information on what to do in an emergency, see page 9.

CAUTION
WHAT TO DO IN AN EMERGENCY

Even if you do your wiring correctly, things can go wrong. Appliances get dropped, furniture moving can damage switches and sockets, vibration can loosen screws, drinks get spilled into TVs and so on. Generally the remedy is simple and obvious, but here are a few essential tips.

■ Electrocution can be fatal, especially if the current passes through a substantial part of the body, and more especially if the casualty is frail or in poor health. If someone is seriously electrocuted they will be unconscious. If they are still connected to the supply, switch it off before touching them or you may end up lying next to them. Phone 999 immediately.

■ There isn't room here for detailed First Aid advice, contact your local St John Ambulance, Red Cross or college for course details. They aren't expensive and many employers will give you a pay rise for becoming a designated first-aider. At the very least you should know the basics of resuscitation and how to put someone in a recovery position. Resuscitation rarely restores breathing, but it can keep someone alive until the paramedics arrive.

■ Many electrical items are made from plastics containing urea which gives off a characteristic smell when hot. If you can smell something reminiscent of the gent's toilet in a busy pub, it's likely you have an electrical fault. If it's not immediately obvious what it is, switch off the main supply immediately before investigating.

■ If you suspect an appliance is faulty, disconnect it and have it checked professionally. If it's something like a cheap toaster it's not worth the expense. If the warranty has expired, dispose of it.

■ If an electrical appliance is on fire **DO NOT** throw water on it. Water can conduct electricity and electrocuting yourself will not help the situation. If you can unplug the appliance safely, then do so; if not, then turn off the main supply before dealing with the fire. Some fire extinguishers are designed for use on electrical fires, but check before you use one.

SAFETY
If someone is seriously electrocuted then call the emergency services immediately.

CHECK
Make sure your fire extinguisher is intended for electrical fires.

1 Knowledge and the Basics

More than 400 years ago Sir Francis Bacon remarked, 'Knowledge is power'. He obviously wasn't thinking about electrical power, but when working with electricity, knowledge is safety. Before we look at some projects you might tackle at home, make sure you understand the basics.

Basic Electrics

Electricity is mysterious stuff, you can't see it, it hasn't got any physical properties like size or colour or weight – but you can certainly feel it if you misuse it. In fact it's a form of energy that flows freely through metal and cannot pass through most other things. This is very convenient as it means we can use wires to carry the current, plastic to insulate the wires from each other (and ourselves) and simply opening a switch or pulling a plug will stop the flow. If electricity doesn't flow, it isn't providing any energy.

In a simple battery powered torch, the current flows out of the + terminal on the battery, through the bulb (or LED) switch (if it's turned on) and back to the – terminal on the battery. That's all very well, but at first glance bears no relation to the sort of circuits used in houses.

The major difference is that there are no '+' and '–' signs on consumer units, plugs, sockets, switches or light fittings because the current we use doesn't just flow in one direction. Batteries deliver Direct Current (DC) which flows steadily like water running downhill from a terminal at a higher voltage to one that is lower. At home (in factories, schools and offices too) we use Alternating Current (AC), which is more like water in a swimming pool with a wave machine – as the level rises and falls the water flows back and forth.

In all domestic AC circuits you will see a Neutral (N) terminal which remains at zero volts (0V) and a Live (L) terminal whose voltage swings up and down from around – 330V to + 330V fifty times every second. The current still flows from high to low, but not steadily, it surges one way then the other as the voltage rises and falls.

When we refer to the UK supply being 230V AC it is actually an alternating current that delivers as much energy as a 230V DC supply would. 230V AC is the 'declared voltage' in the UK, in practice the actual supply is usually around 240V AC, but may vary at peak times.

The frequency of the waves is measured in Hertz (Hz). So, 50Hz just means that the voltage swings from low to high and back to low fifty times per second.

Anatomy of a **Simple Torch Circuit**

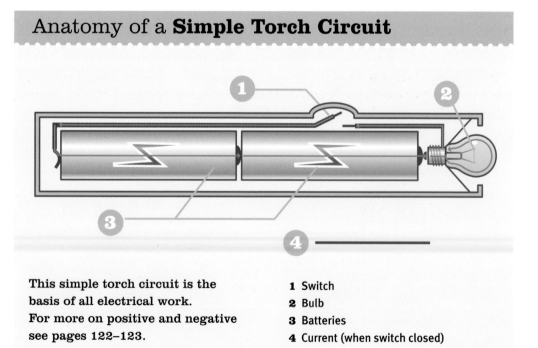

This simple torch circuit is the basis of all electrical work.
For more on positive and negative see pages 122–123.

1 Switch
2 Bulb
3 Batteries
4 Current (when switch closed)

You might notice that many electrical items are marked 220–240V AC 50–60Hz. This is because voltage and frequency vary across Europe, but manufacturers find it easier to make appliances that can tolerate minor differences rather than lots of different models for each country. It also means you can take appliances to other EU countries and expect them to work (with a plug adapter). The US, however, uses 115V.

Conventions

Electric current is actually the movement of electrons from one atom to another along a conductor (see pages 122–123). As electrons carry a negative charge they move in the opposite direction to the conventional flow. This is because electricity was discovered and used long before atomic physics arrived to explain it. The conventions were already in use.

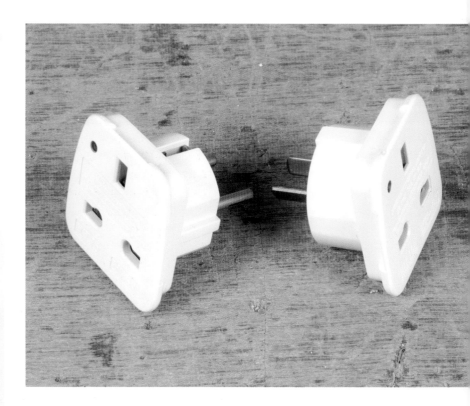

PLUGGED IN
EU countries are standardizing their supplies, but you will still need plug adapters.

Anatomy of **Direct Current – Alternating Current**

Direct current (DC) is like the flow on a water slide. Alternating current (AC) is like the movement caused by a wave machine.

1 Direct Current
2 Alternating Current

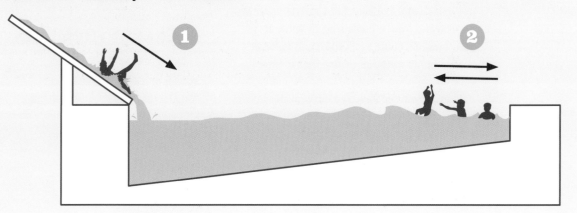

POWER SUPPLY
The national grid
carries power at up
to 400,000V.

Volts, Amps and Watts

The measurements used in electrical work are often misunderstood. 'Voltage' refers to the potential difference between two points on a circuit and is measured in Volts. If you imagine water in a pipe, then voltage is analogous to the pressure that makes it flow.

The amount of flow, the current, is measured in Amps (formally known as Ampères but now abbreviated).

Again, thinking about a water pipe to get more flow you'd have a bigger pipe or increase the pressure to push the water along faster. In electrical terms you can carry more current in a thicker cable or increase the voltage to transmit more energy. The amount of energy transmitted is measured in Watts and a simple formula relates all three measurements:

ENERGY FORMULAS

VOLTS $\times$ AMPS = WATTS
You can turn this around and get:

WATTS $\div$ VOLTS — AMPS
or:

WATTS $\div$ AMPS = VOLTS
For example, a 100W light on a 240V system draws 0.42A

Getting Current to Your House

Unless you live in a very isolated spot, your electricity comes from an electricity company. Their cables are usually underground, but may be strung from a pole to your property, either way it arrives at a meter so the company can check how much current you use, then passes to a Consumer Unit (often called a fusebox, see page 16). Everything before the Consumer Unit belongs to the company and you are not allowed to touch it, the connection from the meter to the Consumer Unit **MUST** be made by someone appointed by the electricity company.

Anatomy of **How Power Reaches Your Home**

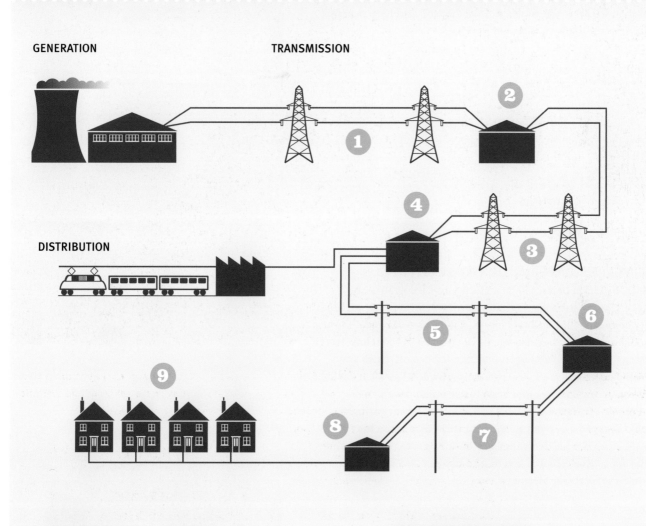

GENERATION

TRANSMISSION

DISTRIBUTION

1 400kV/275kV	**6** Primary
2 Supergrid	**7** 11 kV
3 132kV	**8** Secondary
4 Grid	**9** 415V/240V
5 33kV	

Everything before the Consumer Unit in your house belongs to the supply companies.

Layout of Domestic Wiring

A Modern Consumer Unit (CU)

The Consumer Unit (CU) is where the current coming into your house is split up to supply different devices around the home. At the very least, even in a small flat you will have two circuits, one for lights and another for power sockets. Houses usually have separate circuits for a cooker, shower, heating and possibly one or more for outdoors too for a garage, garden lights and so on. In all but the smallest houses it is normal to have separate circuits for power upstairs and downstairs, and many people have a dedicated circuit in the kitchen for all those gadgets. Some houses also have separate circuits for lights on different floors.

MODERN
A Consumer Unit.

A Miniature Circuit Breaker (MCB)

Each circuit is protected by a Miniature Circuit Breaker (MCB). MCBs are electronic devices that monitor the amount of current flowing in that circuit, if it exceeds the safe maximum either because you have plugged in too many things, or there is a fault, the MCB switches off. Once you have cleared the fault, you can simply switch the MCB back on again, unlike the older style fuses which have to be replaced. MCBs are available in a range of power ratings (also known as current ratings) to suit the circuits they are protecting. Common ratings are 5A for lighting circuits and 30A for power circuits. You might also see 20A or 40A for some radial power circuits.

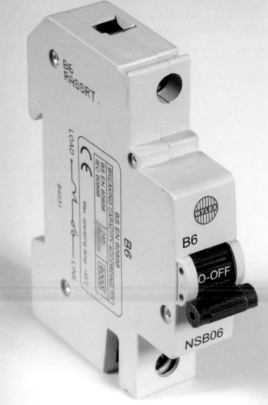

MONITOR
A Miniature
Circuit Breaker.

A Residual Current Device (RCD)

An MCB is designed to protect the circuit against overloads, but it doesn't react to minor faults such as someone electrocuting themselves. If you touch a live circuit the amount of current passing through your body to earth will not cause an overload, but it could well kill you, so in addition to MCBs each CU also has a Residual Current Device (RCD) sometimes called an Earth Leakage Circuit Breaker (ELCB). RCDs work by monitoring and comparing the flow of current on the Live and Neutral lines, they should be identical. If they are not, then current is leaking and the RCD trips.

To prevent unnecessary tripping, some leakage is allowed. High voltage devices such as fluorescent tubes can leak a little current on damp days and heavy duty equipment can cause voltage and current 'spikes' when starting or stopping. Generally, a leakage of 30 milliamps (0.03A) for 40 milliseconds (0.04sec) will trip them. That is still enough to give you a nasty fright though, and could even be fatal, so do not rely on an RCD to protect you. New regulations mean CUs will have twin RCDs, thus protecting all circuits.

TRIP
A Residual Current Device.

✳ Split Load Consumer Units

✔ The RCD is commonly fitted in place of a main switch and if tripped it will switch off everything. This can be annoying at night if all the lights go out.

✔ To counter this, some CUs have a conventional main isolating switch and only connect the power circuits through the RCD. The reasoning is that people are rarely electrocuted by lights as they tend to be on the ceiling and out of reach. It is also quite common to connect a circuit for a freezer in this way so that you don't come home to a pile of rotting food if a trivial fault or power surge trips the RCD while you are away.

✔ To split the load like this you may have to cut the 'bus', that is a metal bar which connects the MCBs to the RCD, or at least remove a section of it. If you buy this sort of CU there will be a fitting guide with details. Some CUs have two RCDs, but this adds substantially to the cost. To meet new wiring regulations, dual RCD split load boards are now being made.

SPLIT
Split Load Consumer Unit.

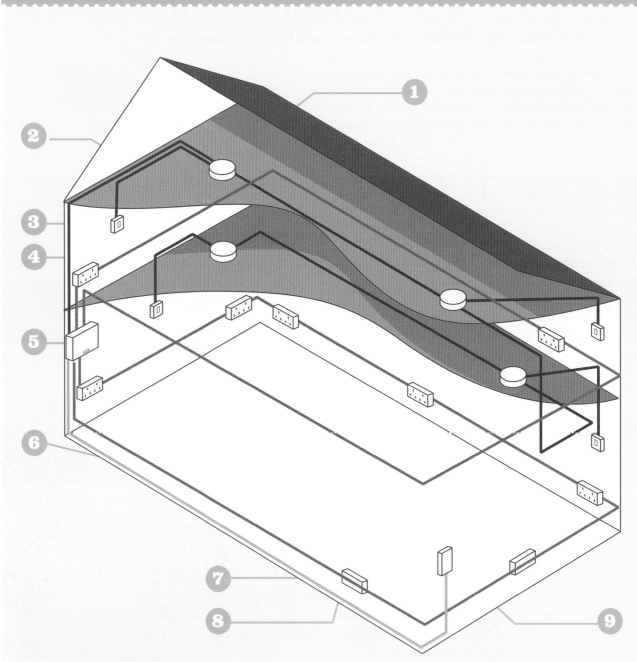

A series of circuits provide
light and power to
your home.

1 Ceiling Rose
2 Lighting Radial Circuit
3 Switch
4 First Floor Ring Final Circuit
5 Consumer Unit

6 Ground Floor Ring Final Circuit
7 Cooker Radial Circuit
8 Fused Connection Unit
9 Cooker Consumer Unit

Fusebox

Prior to 2001 many Consumer Units were fitted with fuses rather than MCBs, giving rise to the common term 'fusebox'. A fuse is nothing more that a thin piece of wire which melts (fuses) when too much current passes through it. Once it has melted no current at all can pass until you replace the fuse wire. This is a lot less convenient than simple switching an MCB back on. Used correctly, fuseboxes are quite safe but many accidents have been caused by people replacing blown fuses with conventional wire, or an incorrect grade of fuse wire.

If you have a common fusebox, such as one of the Wylex range, you could replace the fuse carriers with MCBs (see page 82).

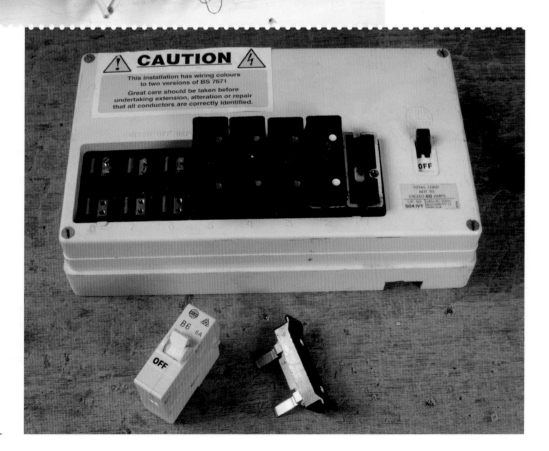

A VERY OLD FUSEBOX
This one needs to be replaced.

A MORE RECENT CONSUMER UNIT
The fuses could be upgraded using MCBs.

Ring Mains

Ring mains (or more properly, Ring Final Circuits) have been common in the UK for more than 50 years. In order to reduce the thickness of the cable required, someone had the bright idea of connecting all the sockets in a continuous ring with both ends connected to the same MCB or fuse in the consumer unit so that current flows through two cables rather than one to reach any point.

The original specification was intended to supply two electric heaters (up to 3kW requiring more than 12A each) with a little surplus capacity for minor appliances. Providing the total load doesn't exceed 30A you can have as many sockets on a ring as you wish, though most people consider 20 to be a reasonable maximum.

Rings are connected using 2.5mm^2 Twin and Earth (T&E) cable (See Cables, from page 34).

Spurs

Sometimes a socket is needed well away from the ring main and it would be a waste of cable to extend the ring to include it. The alternative is a spur, that is a branch off the ring using a single piece of cable. It is conventional to connect the spur to a socket on the ring, though in some cases it might be preferable to use a junction box.

Anatomy of a **Basic Ring Circuit**

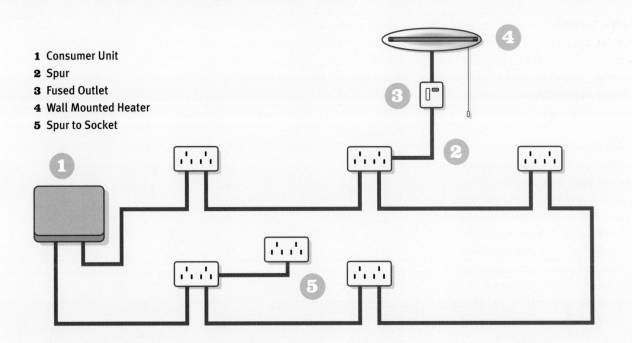

1 Consumer Unit
2 Spur
3 Fused Outlet
4 Wall Mounted Heater
5 Spur to Socket

Note, 2.5mm² T&E is rated at 20A so cannot be used to supply two 13A loads, though one double outlet is allowed on a spur. Spurs are a useful way of increasing the number of sockets on an existing system but are considered bad practice on a new circuit.

Radial Final Circuits

The alternative to a ring circuit is a radial circuit where a series of single cables radiate out from the CU to the sockets or appliances. This is rare in the UK for normal sockets but is used to supply appliance that require more than 13A, such as immersion heaters, cookers and electric showers. Each appliance is connected to a dedicated MCB in the CU using cable appropriate to the current it consumes. (See Cables, page 34.)

Lighting Radial Circuits

The load of a lighting circuit is much less than on a power circuit, easily within the capacity of a single run of 1mm² or 1.5mm² cable, so a radial final circuit is used from the CU.

SIMPLICITY
No matter how sophisticated your lighting, the circuitry is quite simple.

6/2163707

The layout shown here is for a house with a suspended wooden floor with the cables running underneath it. In many modern houses the floor at ground level, where the kitchen is most likely to be, would be solid. In this case, the cables would come down from the ceiling.

■ Power

RING MAIN In modern systems a cable loops from the consumer unit to each socket or fused connection unit (FCU) in turn, then back to the same fuse at the consumer unit in the form of a ring. Power can flow from each side of the ring, thereby supplying double the normal current value of a single cable. The cable is 2.5mm² twin and earth, protected by a 30A fuse (or 32A MCB).
A ring main can supply an unlimited number of sockets and FCUs to a maximum load of about 7,000W, providing it does not serve a floor area of more than 100m².
The maximum length for a ring final circuit should be 75m.

RADIAL The power circuit cable does not have to return to the consumer unit when the floor area is no greater than 50m². However, the circuit fuse at the consumer unit must be restricted to 20A.

NOTE The maximum size appliance which may be connected to the power circuit (whether ring or radial) is 3kW.

– – – Ring Main Spur

One branch or spur can be taken from the terminals of each socket to supply a new socket. Alternatively, the spur could be taken from a junction box wired into the ring main.

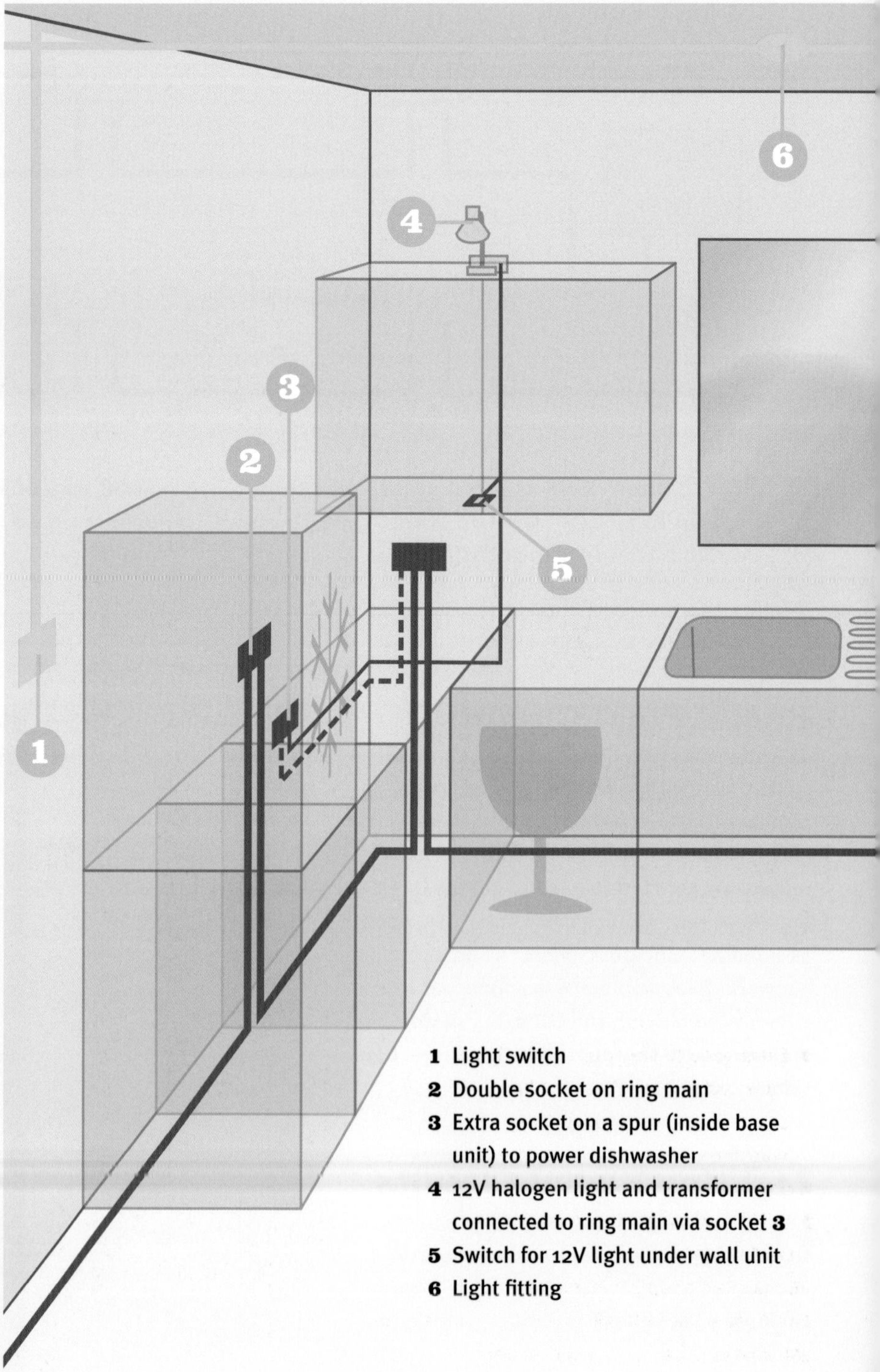

1 Light switch
2 Double socket on ring main
3 Extra socket on a spur (inside base unit) to power dishwasher
4 12V halogen light and transformer connected to ring main via socket 3
5 Switch for 12V light under wall unit
6 Light fitting

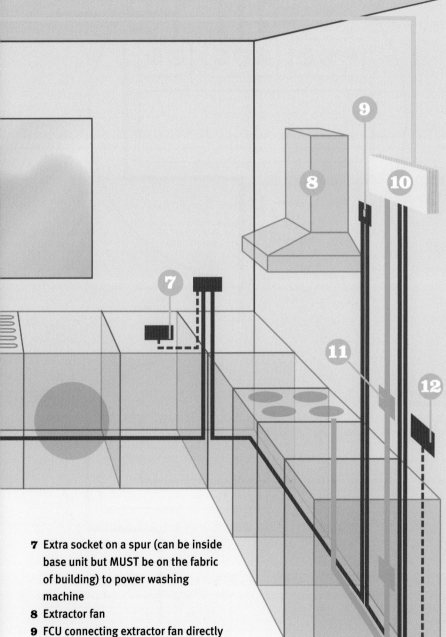

Lighting

One cable is looped from one lighting point to the next, usually with 1.5mm² twin and earth cable. The circuit is protected by a 5A fuse (6A MCB) at the consumer unit.

FUSED CONNECTION UNIT (FCU)

An FCU that incorporates a switch can be used to connect kitchen appliances, such as an extractor fan, directly into the ring main without the need for a plug socket. An FCU can be used to supply one or more sockets from the ring main providing the current demand does not exceed 13A and at least 1.5mm² twin and earth is used as the spur cable.

Cooker Radial

A dedicated circuit is usually provided to any single appliance with a high current rating, typically the cooker. All-electric hobs and double ovens come within this category. Usually, 6mm² twin and earth cable is used, protected by a 30A fuse (32A MCB) and this is normally adequate for appliances up to 17.6kW.

NOTE Single electric ovens (rated no more than 3kW) can be connected to the power circuit, but must be protected by a 13A fuse.

7 Extra socket on a spur (can be inside base unit but MUST be on the fabric of building) to power washing machine

8 Extractor fan

9 FCU connecting extractor fan directly to the ring main

10 Consumer unit (in adjacent room)

11 Cooker control unit

12 Extra socket from junction box

Lighting circuits are more complicated than ring mains because you need to install switches. If you have conventional ceiling roses then there are terminals in the rose to connect to a switch (or two in the case of two-way switching). Wall lights and some fluorescent fittings do not have roses so you may need to use a junction box.

Inside a ceiling rose there are three terminal blocks, one with two holes, the others with three. Wired normally, there is one cable per hole as shown opposite. There is also an earth connection.

CEILING ROSE
CONNECTIONS

1 Radial Circuit
2 Radial Circuit

3 Switch Circuit
4 Earth Connection
5 To Light Fitting

TWO (OR MORE) LIGHTS CAN BE WIRED IN PARALLEL

If you want two or more lights to be operated by the same switch, the second (third etc) are connected in parallel to the first.

TWO-WAY SWITCHING IS VERY SIMPLE
Two-way switching is used on stairs so you can switch the lights on or off from either the top or bottom. The ceiling rose is connected in much the same way, but instead of a T&E cable to a single switch, two single cables are used connecting to the common terminals of two two-way switches. The other terminals (L1 and L2) on the switches are connected with T&E.

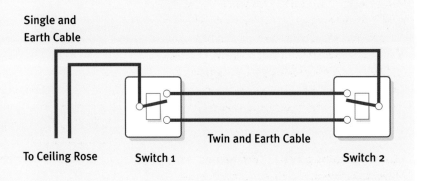

Single and
Earth Cable

To Ceiling Rose   Switch 1   Twin and Earth Cable   Switch 2

EVEN THREE-WAY ISN'T THAT HARD . . .
Three-way switching is also possible, and useful in hallways where you might want a switch by the front door, kitchen door and lounge door. To do this you need two two-way switches and one intermediate switch. Intermediate switches have four terminals and are arranged so that throwing the switch swaps the connections over. In one position terminal 1 is connected to 2 and terminal 3 to 4, flip the switch and 1 is connected to 4 and 2 to 3. In use, the intermediate switch is connected between the two-way switches as shown.

To Ceiling Rose

Switch 1   Switch 2   Switch 3
              Intermediate

To Ceiling Rose

Fittings without Ceiling Roses

Some light fittings, especially fluorescent strip lights and wall lights, are not designed to hang from ceiling roses, but you still need to make all the same connections. The usual solution is a small (5A) junction box. These are normally hidden from view in the loft space, between the ceiling and the floor above, inside a partition or sometimes behind the skirting board.

A 5A JUNCTION BOX SUITABLE FOR LIGHTING CIRCUITS
Junction boxes have four terminals which are unlabelled, you can use them as you wish. There are cut outs in the sides and the lid is designed so that it can reveal either 1, 2, 3 or 4 cut outs.

Earthing

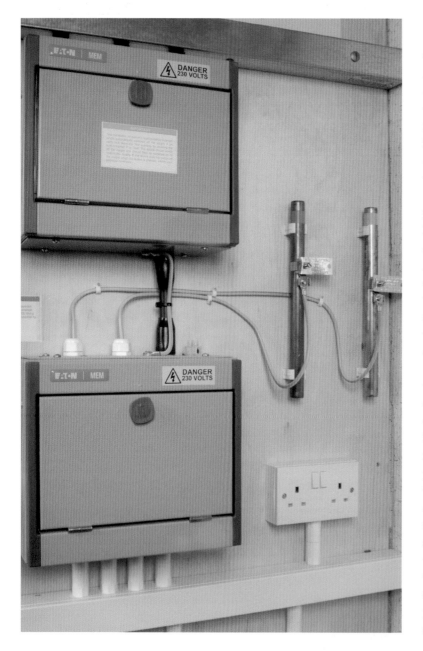

EARTH
Correct earthing is
absolutely essential.

To appreciate how important earthing is, just think what might happen without it. If a live wire comes loose inside an appliance and touches a metal case, the case becomes live. You touch it and at the very least you get an unpleasant shock. If the casing is earthed, current flows from Live to Earth, greatly exceeding the rating of the fuse or MCB which blows, cutting off the supply.

Earthing only works if it is done properly. If one ceiling rose on a radial lighting circuit isn't earthed then neither are any of the subsequent ones, and in case getting a shock from a plastic light fitting on the ceiling seems unlikely, don't forget that light switches have metal screws, and some even have metal fascias. All earth wires should ultimately lead back to the earth connection of the CU.

Earthing around plumbing is even more vital. Copper pipes are extremely good conductors of electricity but some of the connectors used in plumbing work can have a high electrical resistance, especially if they are a little corroded. The use of plastic pipes and fittings has exacerbated this problem by leaving some metal fittings electrically isolated from the rest of the system. A fault in a boiler, water heater, shower, central heating pump or motorized valve could cause a pipe to have a high voltage whilst another adjacent one is at a lower voltage or even earthed. This potential difference is potentially fatal to anyone who touches both pipes.

To avoid problems, all metal pipework and plumbing should be connected together using earthing straps and earthing cable, a system known as Equipotential Bonding. The minimum cable size is generally considered to be 4mm² but many people use heavier. The standard clamps are metal and screwed on and look rather industrial. Ezybond clamps are plastic, quicker and easier to fit and look neater. Whatever sort of clamp you use it should be clearly labelled as: **Safety Electrical Connection – Do Not Remove.**

If you have old lead pipes, corrosion on the surface can prevent earthing clamps making a good contact. Such pipes should be bonded by a professional who can solder a connection to the pipe. Better still, replace the lead pipes. Lead can leach into the water and is poisonous.

CAUTION
LABELS
'Do Not Remove'
warning labels
are essential.

EARTHING CLAMPS AND EARTH CABLE

EARTHING CLAMPS

1 Double row of interlocking teeth
2 Safety Lock
3 Grip
4 Suitable for earth wire sizes A–C
5 Pipe Size
6 Guide Vanes
7 Conductor

EARTHING CABLE

Basic Tools

omestic electrical work doesn't require many specialist tools, or any special skill. Mostly it's a matter of knowing what you are doing, doing it methodically, and checking everything you have done. Then checking it again. You will need:

Screwdrivers

A selection of screwdrivers, the usual #2 Posidriv and a medium-sized flat bladed screwdriver for fixing sockets and the like, together with smaller cross point and flat for terminal screws. A circuit test screwdriver is useful to make sure that a circuit you are working on really is isolated.

DO NOT try to use it as a diagnostic tool by leaving the power on and poking around, it's dangerous and there are better ways of finding faults.

Pliers

Combination pliers can be used for cutting and stripping cable and are useful for other tasks too (such as holding and turning things), but general purpose tools are never as good as tools made for specific tasks.

If you plan on doing anything more than replacing the odd switch, then side cutting pliers and wire strippers will make life a whole lot easier.

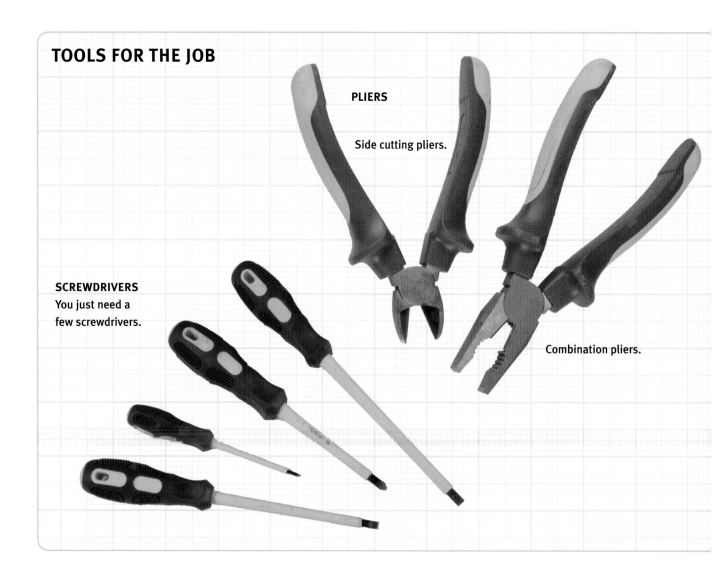

TOOLS FOR THE JOB

PLIERS

Side cutting pliers.

Combination pliers.

SCREWDRIVERS
You just need a few screwdrivers.

Most pliers have insulated handles as a safety precaution, but this isn't an invitation to try working on live cables! Always ensure the current is off before cutting or stripping wires.

Wire and Pipe Detector

If plumbers and electricians follow correct procedures then pipes and cables should either be buried deep out of harm's way, protected by steel conduit or guarding, or run in predictable paths which you can avoid. However, in anything but a new house there is always a chance that some cowboy has been there before you, so before you start drilling into the wall or nailing down loose floorboards it's a good idea to check what lies behind or under.

Variously known as Wall Scanner, MultiDetector or Stud, Wire and Pipe detectors, there are dozens to choose from. Reviews suggest that the basic metal detection types are best for finding hidden cables and pipes, those that attempt to locate wooden studding and other materials are less reliable.

Multimeter

Also known to old hands as an AVO meter (because it measures Amps, Volts and Ohms). You don't need one if you are only intending to add a couple of sockets or move a light switch, but they are useful for testing new installations and very good for fault finding.

toptip*
Cheap pliers, wire cutters and strippers are a waste of money. They are made from low-quality steel and go blunt very quickly.

WIRE STRIPPERS
Wire strippers are useful but not essential.

DETECTOR
A cable and pipe detector can prevent disasters.

SS-1977
3-IN-1 METAL/VOLTAGE/STUD DETECTOR

MULTIMETER
A multimeter is very useful for testing and fault finding.

General DIY Tools

USEFUL TOOLS

The following are not specifically electricians' tools, but are useful nonetheless:

CLAW HAMMER
A hammer to hit it with is essential too.

BOLSTER
Chasing out grooves in plastered walls or cutting spaces for switch and socket boxes is much easier with a brick bolster than a narrow cold chisel.

WALL BOARD SAW
Sometimes known as a Pad Saw or Dry Wall Saw, these are useful for cutting holes in partition walls for switches and socket boxes.

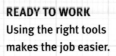

READY TO WORK
Using the right tools makes the job easier.

MEASURING

TAPE MEASURE AND SPIRIT LEVEL
You will need a tape measure and spirit level to set out cable runs and position switches and sockets.

BOX TEMPLATE
A box template with a built-in spirit level and a height gauge to set sockets all the same height is a time saver.

DRILL
Unless all your electrical work is on partition walls, or you are very keen on hard work, you'll need some sort of drill to Rawlplug socket boxes. They are useful for chain drilling chases in plastered walls too. A basic model will do, but a hammer drill is useful for tackling brick or block walls.

PLASTERER'S TROWEL
And finally a plasterer's trowel is useful to make good all the plasterwork you've hacked out.

Cables

SWITCHED ON
Assorted cables and fittings.

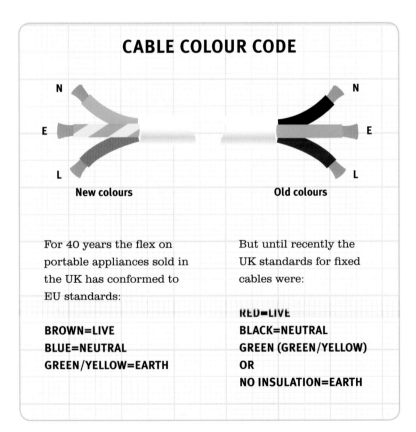

CABLE COLOUR CODE

New colours   Old colours

For 40 years the flex on portable appliances sold in the UK has conformed to EU standards:

BROWN=LIVE
BLUE=NEUTRAL
GREEN/YELLOW=EARTH

But until recently the UK standards for fixed cables were:

RED=LIVE
BLACK=NEUTRAL
GREEN (GREEN/YELLOW)
OR
NO INSULATION=EARTH

At first sight the range of cable types, thicknesses and colours seems daunting, but it's really quite logical.

Wire and Cable

This is the first confusion, should you ask for wire or cable? The difference is very simple; a wire has a single conductor, a cable is made of two or more wires. Some wire is multi-stranded, it is made of lots of fine wires rather than one thicker one, but they are all inside the same insulator carrying the same current. The wires in a cable may be single core or multi-stranded. Almost all fixed wiring uses cable, even the links from a ceiling rose to a two-way switch are Single and Earth so have two wires. The exception is Earth wire used for bonding pipework and fixed appliances.

Current Carrying Capacity

The amount of current a wire can carry depends on its cross-sectional area, which is why you see cables referred to as $1mm^2$, $2.5mm^2$ and so on. In the case of multi-stranded wires the measurement refers to the total area of all the conductors. Length has some bearing on capacity too, but it isn't normally an issue in domestic wiring. Cabling a large office or factory is a different matter. See the tables on pages 33 and 35 for guidance on choosing the correct cable for various circuits.

Cable Colours

The plastic sleeving around wire has two functions, electrical insulation and colour coding that tells you which wire is connected to what.

SOCKETS AND CABLES

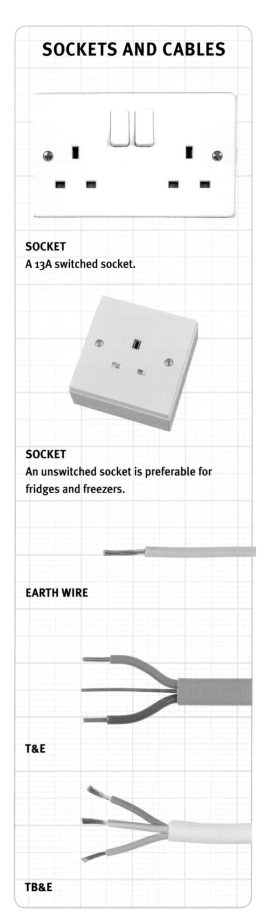

SOCKET
A 13A switched socket.

SOCKET
An unswitched socket is preferable for fridges and freezers.

EARTH WIRE

T&E

TB&E

If you buy cable today it will conform to BS 6004 and use the European colours, but you may be working with or extending and modifying installations done some time ago with the older style cable. Take care when mixing them.

Conductors running to and from switches are both live (when the switch is on) so both should be brown. You can buy T&E with two brown wires, but it's not economical for the occasional job. You can use conventional Brown/Blue T&E but you MUST use brown sleeving over the ends of the wires inside the switch and ceiling rose.

Flex

Flexible cable (flex for short) is used in portable appliances and even fixed things such as electric heaters where the cable from the fused connection point is exposed and may be moved. Pendant lamp holders hang from ceiling roses on flex. The wires in flex are very fine multi-stranded so they bend easily and the plastic used to insulate them is also chosen for its flexibility and resistance to cracking when flexed.

POWER POINT
Sockets to suit your decor.

Flex Cable Selection

Flex can be damaged, and if it is the appliance should not be used until the flex is replaced. Many modern appliances are made of plastic and constructed in such a way that the user never comes into contact with any metal parts that could become live. Such devices are referred to as double insulated and can be connected using two-core flex. Don't assume it applies to all appliances though. If the original cable was three-core, you should replace it with the same, and make sure the earth connection is sound.

Conductor Size	Current	Maximum Power	Typical Use
0.5 mm²	3A	800W	Lighting
0.75 mm²	6A	1600W	Lighting
1.0 mm²	8A	2300W	Lighting
1.25 mm²	13A	2990W	Power
1.5 mm²	15A	2450W	Power
2.5mm²	20A	4600W	Power

SOCKETS AND CABLES

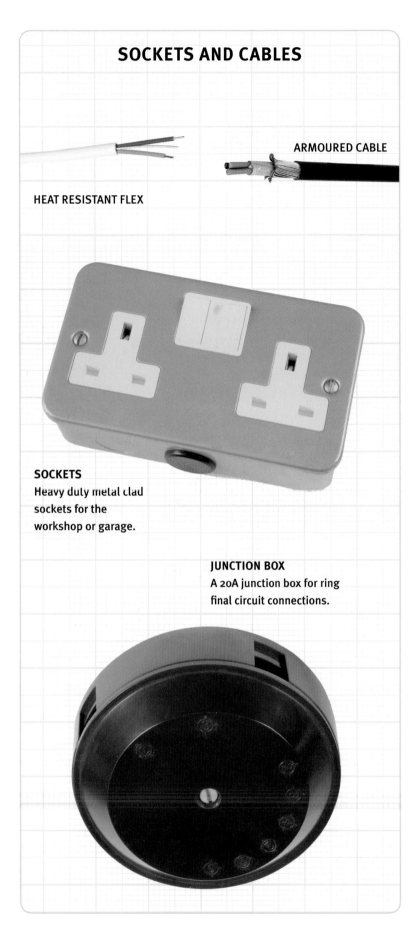

ARMOURED CABLE

HEAT RESISTANT FLEX

SOCKETS
Heavy duty metal clad sockets for the workshop or garage.

JUNCTION BOX
A 20A junction box for ring final circuit connections.

Heat Resistant Flex

Some things can reasonably be expected to get hot in use (soldering irons for example) and these are often fitted with heat resistant flex. If you work with hot tools and materials it's a good idea to replace all flexes with a heat resistant type. Heavy duty heat resistant flex is used to connect immersion heaters, it is proof against water at 90°C so that a plumbing problem doesn't also become an electrical problem!

Cable

When electricians refer to just 'cable' they normally mean the stuff installed in buildings, as opposed to 'flex' for appliances or 'armoured' for outdoor or industrial use. The most common form is Twin and Earth (T&E) with two insulated conductors for Live and Neutral and a bare earth wire, all in a PVC outer sleeve. Wherever the earth cable is exposed, inside socket boxes or behind light switches for example, green and yellow earth sleeving should be fitted over the bare wire.

Up to 2.5mm² solid single core wires are used in T&E, beyond that a solid core would be awkward to handle and difficult to bend round corners so stranded cables are used.

toptip*

If you are likely to do more than one small wiring job, buy cable on reels. It's cheaper in the long run and you waste less.

Armoured Cable

If you are running cable outside (for example to a shed or for garden lighting) you can either use conventional cable inside a conduit, or armoured cable. Armoured cable is obligatory if it is going to run underground. The armour consists of a tough outer sheath surrounding steel strands. Inside that is a layer of softer plastic and then the actual wires.

Armoured cable is mostly used in industrial situations where three phase supplies are the norm. Three phase is outside the scope of this book but you should note that the phase conductors in new cable are coded brown, black and grey. When used for single phase wiring you should use:

COLOUR CODING FOR ARMOURED CABLE

BROWN=LIVE
BLACK=EARTH
GREY=NEUTRAL

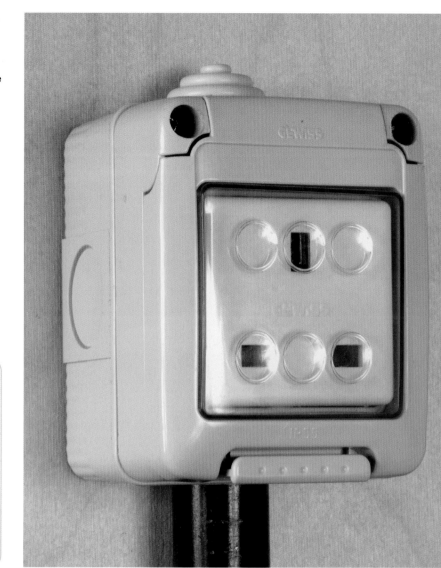

SAFE
Waterproof sockets are absolutely essential outdoors.

Fixed Cable Selection

Conductor Size	Current	Maximum Power	Typical Use
1.0 mm²	10A	2.4kW	Lighting
1.25 mm²	13A	3.12kW	Lighting
1.5 mm²	15A	3.6kW	Lighting
2.5 mm²	20A	4.8kW	Power
4.0 mm²	25A	6kW	Cooker
6.0 mm²	45A	10.8kW	Cooker/Shower

Current handling capacity also depends on the length of the cable. For long runs and high current, for example a 10.8kW shower a long way from the Consumer Unit should use 10mm² cable. Cable sizes need to be adjusted for thermal installation and cable routes, professional advice should be sought if you are not sure.

Sockets and Connections

SOCKETS

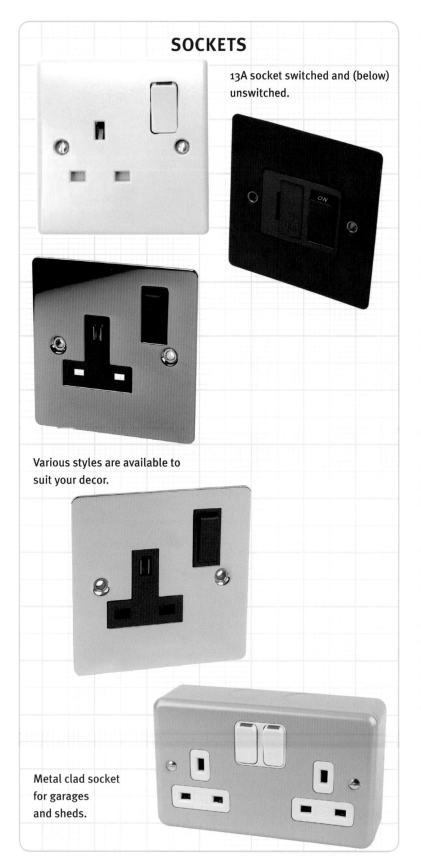

13A socket switched and (below) unswitched.

Various styles are available to suit your decor.

Metal clad socket for garages and sheds.

The domestic sockets and plugs used in the UK and Ireland (and some former colonies) were designed to suit our ring main systems. With a circuit capable of delivering up to 30A it takes a very serious fault to blow the consumer unit fuse so it was decided that each plug should carry its own fuse to suit the appliance in question. Most other countries have lower rated circuits, in some cases a separate fuse or MCB for every socket, so plugs can be simpler and smaller.

Many modern appliances do not require an Earth connection, but all 13A plugs have to have an Earth pin because it is used to open the shutters that cover the Live and Neutral sockets. Some low voltage power supply units such as mobile phone chargers have a plastic pin for this purpose.

Most 13A sockets now incorporate switches, but these are just for convenience not a legal requirement. For appliances such as freezers (which are rarely turned off), you may prefer an unswitched socket.

Sockets are made in a variety of materials and finishes. White plastic is still common because it is cheap and looks reasonably good with any decor; modern designs are slimmer, slightly curved and look less heavy than the older styles. Various retro styles in brass are available and recently there has been a move towards sleek metal designs with hidden screws.

For garages and sheds the industrial style metal clad sockets are useful as they can withstand a certain amount of rough treatment. All interior sockets are designed to meet the same standard, BS 1363, and are interchangeable. For outdoor use, weatherproof designs are available.

Fused Connection Units

Where an appliance such as a heater is fixed, rather than portable, a Fused Connection Unit is normally used in place of a socket. These are also rated at 13A and use the same replaceable ceramic cartridge fuses as standard plugs.

The simplest variety has a switch, a fuse and a flex outlet and is commonly used in a kitchen to connect a water heater or extractor hood. If the appliance is out of reach, such as a radiant heater high up on a wall, then the normal practice is to use a connection unit without a flex outlet mounted in a convenient location, combined with a separate flex outlet close to the appliance. This sort of arrangement is referred to as a 'switched spur'.

Conventional 13A sockets should not be used in a bathroom as fiddling with a plug with wet hands whilst standing in a puddle of water is a sure-fire way of electrocuting yourself. Even fused outlets are prohibited because the fuse holder could be opened (or the switch operated on switched versions). The basic rule is that no source of electrical supply should be accessible from a position where one can be in contact with water at the same time.

Junction Boxes

Wherever possible connections to ring final circuits are made utilizing an existing socket, but sometimes you might wish to add a spur to the ring where there is no convenient socket; in this case a junction box is needed. Junction boxes for power circuits are rated at 20A. Do not be tempted to use a smaller 5A one, these are only suitable for lighting circuits.

Electric radiators and heated towel rails should be connected to a switched spur with either a conventional switched fused connection unit outside the bathroom, or an unswitched FCU combined with a 'pull switch' mounted on the ceiling (sometimes high on the wall, but must be out of reach). Such switches must be in accordance with BS 3676, and this is normally indicated on the label. The current rating of the switch must be suitable for the load, and unless the device has a 'power on' indicator then a switch with a neon is a good idea.

Immersion heaters should be connected to a radial final circuit through a double pole switch. Double Pole (DP) switches are actually two normal single pole switches side by side in the same package and joined together so they operate simultaneously. They have four terminals arranged in two pairs and are commonly used to switch off both the Live and Neutral. Used like this they are known as isolating switches.

It may seem unnecessary to cut off the neutral connection as well as live, but there is a sound safety reason. If there is a fault on the neutral wire somewhere between an appliance and the Consumer Unit, and the appliance is switched on, the neutral wire can become live (current passes through the appliance, but cannot reach the CU via the broken connection.) This is rare, especially on ring circuits where two neutral wires would have to be broken or disconnected, but it can happen. It is even more likely on a radial circuit for something like a shower or immersion heater, and with all that well earthed plumbing around the chance of a fatal accident is that much higher. Hence the isolating switch. A fused connection unit isn't necessary as the radial circuit is connected to a dedicated MCB.

Flex outlets are safe to use in bathrooms, providing the switch is outside.

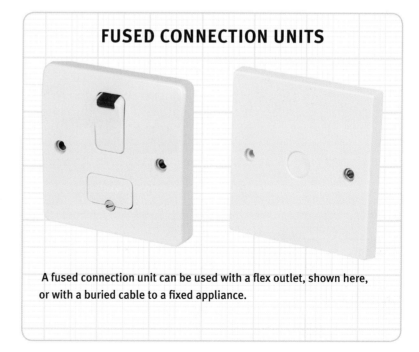

FUSED CONNECTION UNITS

A fused connection unit can be used with a flex outlet, shown here, or with a buried cable to a fixed appliance.

SWITCHES AND SOCKETS

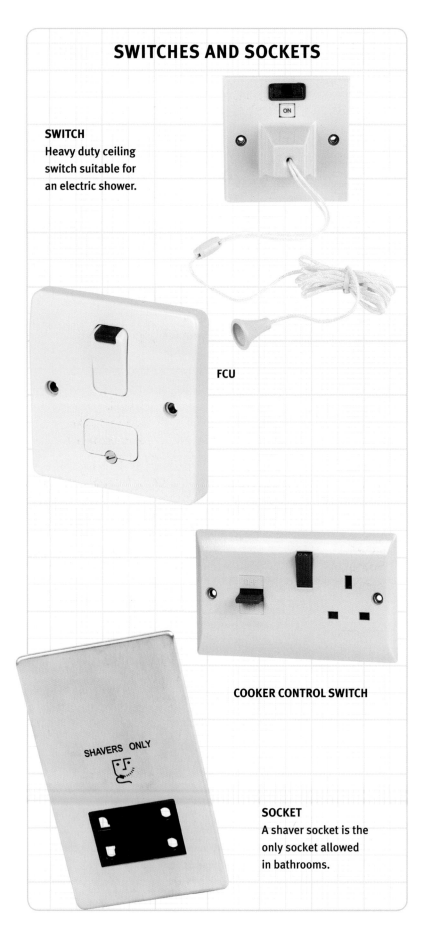

SWITCH
Heavy duty ceiling switch suitable for an electric shower.

FCU

COOKER CONTROL SWITCH

SHAVERS ONLY

SOCKET
A shaver socket is the only socket allowed in bathrooms.

Electrical Shower Switches

Electric Showers usually use more than 13A. They are also connected via a radial final circuit, sometimes using 2.5mm² cable rated at 20A but sometimes thicker cables carrying far higher currents are used. Showers must be connected via double pole switches with a suitable rating, the ceiling mounted design shown can handle up to 45A. No flex outlet is needed because showers are designed to be connected direct to the cable in the wall, and no FCU either because, as with immersion heaters, the shower has its own MCB.

Cooking Controls

Electric cookers can consume enormous amounts of power, so the cable and control points have to be heavy duty. A 6mm² cable is used and a control point rated at 45A together with an equally heavy-duty outlet point which is usually mounted under the work surface or behind the cooker. Cooker control points incorporate a 13A socket, which seems to be a hangover from the days when no one had enough sockets in their kitchen. Specially deep pattresses or steel boxes are needed to mount cooker controls to accommodate the 6mm² cable.

Shaver Sockets

Shaver sockets are an exception to the 'no sockets in the bathroom' rule. These special sockets are isolated from the mains supply, drawing their current from a small transformer. The output is limited to 20VA (Volt Amps), if you plug in anything drawing more power the socket will turn off. Many are dual voltage, 230V and 115V to suit a range of shavers and electric toothbrushes. Because shaver sockets draw so little current they may be connected to a lighting circuit if it is more convenient than connecting to the ring main.

Socket Boxes

Sockets are mounted on steel or plastic boxes attached to the wall. Steel boxes are used on plastered walls where they can be recessed into the plaster. Steel boxes are available to suit single or 'two gang' or 'three gang' (double or triple) sockets and come in a variety of depths. The shallowest, 16mm, are only suitable for light switches. The heavier cables used in ring final circuits need 35mm boxes though you can use a 25mm box for a fused spur with only a single cable coming into the box. Some special sockets, for example cooker control points or shaver sockets, require even deeper boxes, check when you purchase the socket or connection point.

Plastic surface mounting boxes are used on solid walls, and can be useful if you don't want to spoil the finish on a wall when adding a new socket, for example putting a shaver point in a bathroom with tiled walls.

Dry lining (or plasterboard) boxes are also made of plastic but are designed to fit into plasterboard partition walls. The boxes have lugs that slide into place behind the board and are anchored by the same screws that fix the socket in place. They also have a thin flange to cover any rough edges around the hole, making for a very quick and tidy job. Dry lining boxes are 35mm deep for either switches or sockets.

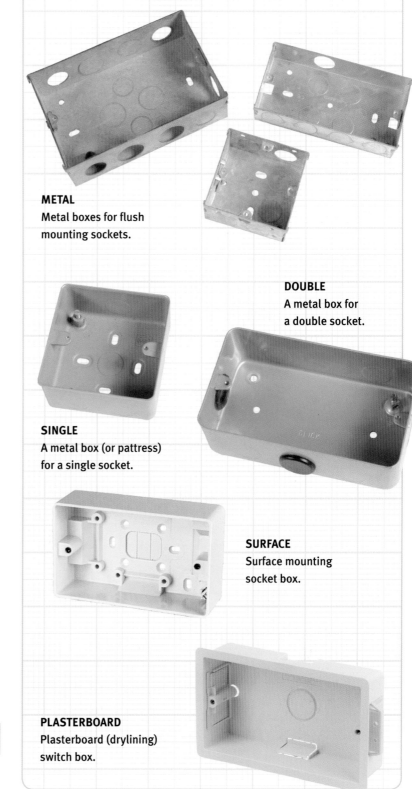

SOCKET BOXES

METAL
Metal boxes for flush mounting sockets.

DOUBLE
A metal box for a double socket.

SINGLE
A metal box (or pattress) for a single socket.

SURFACE
Surface mounting socket box.

PLASTERBOARD
Plasterboard (drylining) switch box.

toptip*

If you have space, fit deeper boxes. It is so much easier to fit all the wires in!

Light Fittings

Ceiling Roses

A ceiling rose is a specialized form of junction box optimized for connecting into a radial final circuit, with terminals for a switch and for the flex leading to the bulb holder. They are generally round, white and made of plastic, which suits most ceilings, though more exotic colours and materials are available.

Pendant Lamp Holders

The standard UK bayonet cap (BC) lamp holder designed to hang from a ceiling rose. It is worth considering buying roses and pendants ready wired, they work out a few pennies more expensive but save a lot of time and fiddly work.

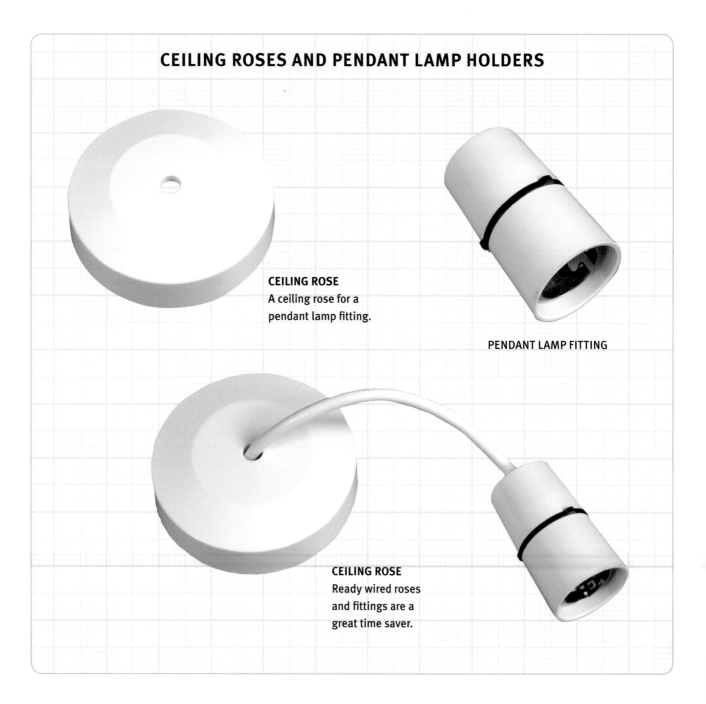

CEILING ROSES AND PENDANT LAMP HOLDERS

CEILING ROSE
A ceiling rose for a pendant lamp fitting.

PENDANT LAMP FITTING

CEILING ROSE
Ready wired roses and fittings are a great time saver.

Other Lights

There is a huge variety of other light fittings; some for ceilings, some for walls; some close fitting, some recessed, some dangling from hooks or brackets. What they have in common is they don't attach to a ceiling rose. Some of the ceiling designs have a 'boss' that looks like a rose, but you rarely find the same connections inside. Most have just a simple piece of connecting block with L and N terminals and a screw on the lamp body for the earth connection. If there is room you might be able to replace the connector with a longer strip with four terminals and wire it like a rose. If that isn't possible then you should use a junction box.

JUNCTION BOXES

Junction boxes for lighting circuits are similar to, but smaller than, those used in power circuits. They are usually hidden in lofts or under floors.

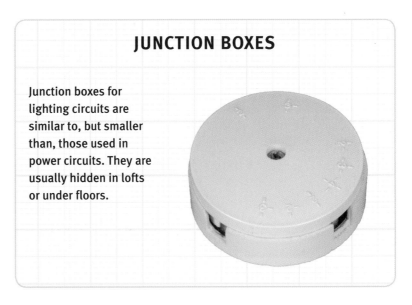

LIGHTS

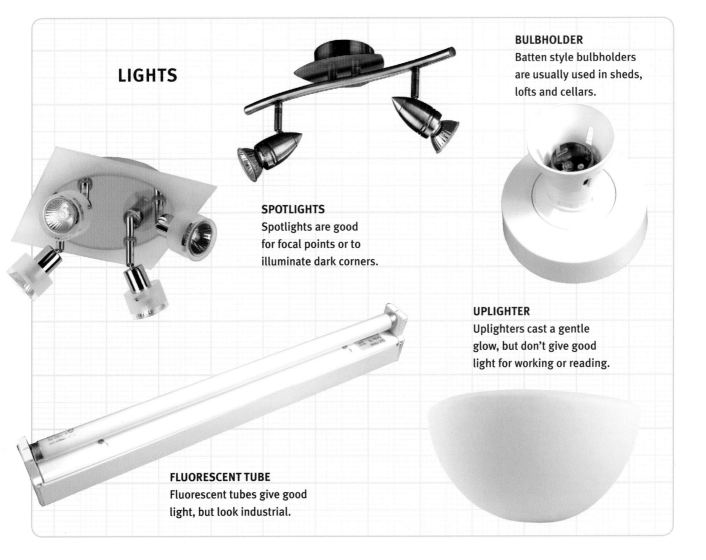

SPOTLIGHTS
Spotlights are good for focal points or to illuminate dark corners.

BULBHOLDER
Batten style bulbholders are usually used in sheds, lofts and cellars.

UPLIGHTER
Uplighters cast a gentle glow, but don't give good light for working or reading.

FLUORESCENT TUBE
Fluorescent tubes give good light, but look industrial.

Switches

SWITCH
A basic light switch.

Anatomy of **Circuit Symbols**

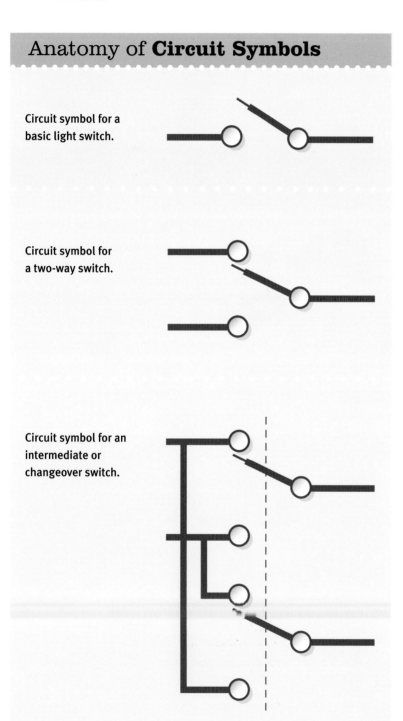

Circuit symbol for a
basic light switch.

Circuit symbol for
a two-way switch.

Circuit symbol for an
intermediate or
changeover switch.

At its simplest, a switch is just a mechanism for interrupting the flow of current. There are two terminals and a metal contact which opens and closes. It doesn't matter which way round you connect them.

Slightly more complex is the two-way switch (Double Throw). These have three terminals labelled C (common), L1 and L2. There is no 'off' position with a two-way switch, current can flow between C and L1 or C and L2, flicking the switch just diverts it from one to the other.

Making a two-way switch is scarcely more complicated than a simple on-off switch, so prices are similar. In fact, prices are so similar that some stockists don't keep on-off types at all and some budget makers don't even make them. A two-way switch can be used as a simple on-off switch by simply using C and L1 and ignoring L2.

An intermediate switch has four terminals, but actually consists of two two-way switches in a double-pole configuration. Normally this would have six terminals but to create a changeover operation four of the terminals are internally connected in pairs. In operation, flicking the switch swaps the connections so that the options are either terminal 1 connected to 3 and terminal 2 to 4, or 1 to 4 and 2 to 3.Most switches are designed to fit the same size boxes as sockets and come in a matching range of styles and colours. Multiple switches on the same plate are referred to as 'gangs'. Two-gang and three-gang just means you get two or three switches in a single fitting. A single switch will fit on a 16mm deep box, but a three-gang of two-way switches needs a lot of cables so a deeper box is advisable. If a hard wall makes fitting a deep box difficult, consider using more boxes and one or two-gang switches instead.

Architrave Switches

Rather than cut into the wall you might prefer to fit architrave switches. These are narrower and designed to fit in door surrounds. In partition walls it is sometimes possible to feed the cable inside the wall from above and out though a hole in the door frame without damaging the wall at all.

Ceiling Switches

Sometimes called pull cord switches, these are essential in bathrooms. They used to be common in bedrooms hanging over the bed, but bedside lights and wall lights have made them obsolete there.

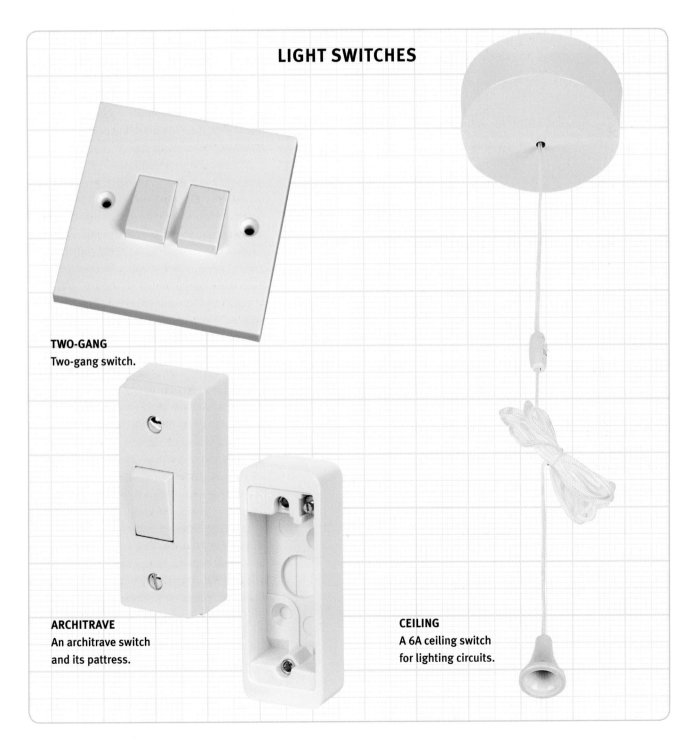

LIGHT SWITCHES

TWO-GANG
Two-gang switch.

ARCHITRAVE
An architrave switch and its pattress.

CEILING
A 6A ceiling switch for lighting circuits.

Dimmer Switches

Remember the wave analogy for alternating current on page 13? Good, because it helps you understand dimmers. The circuitry in a dimmer reacts to the rising and falling voltage level; at 0 volts it switches off and remains off until the voltage rises to a given level (set by you operating the control). It cuts off again when the wave falls to zero, then switches on when the same level below zero is reached. The result is that instead of a smoothly changing alternating current, you get a hundred pulses every second (50 in each direction on a 50Hz supply). This is faster than the filaments in bulbs react, so rather than seeing pulses we just see less light, with longer pulses making it brighter and shorter ones dimmer.

This works well with incandescent lamps but not with standard fluorescent tubes, they just can't switch on and off that fast. More recent fluorescent fittings are designed to work with dimmers. Some models rely on four wires to the lamp fitting, two providing full power and two from the dimmer which connect to a circuit in the lamp fitting that controls the light output. Newer models are able to operate with the normal two wires. If you are buying a new fluorescent, check the label and follow the manufacturer's connection instructions carefully.

At the moment most low energy lamps cannot be dimmed; however. the government is moving towards banning the sale of conventional incandescent lamps and the industry is working hard to produce dimmable LE (low energy) lamps at reasonable cost.

In most cases a dimmer switch can simply replace a conventional switch, but do check the current rating carefully. Most are suitable for up to 500W but may be de-rated to 400W or less for fluorescents.

LOW LIGHTS
Future homes could have dimmable low energy lights.

Outdoor Switches

As with sockets, there are light fittings and switches designed for outdoor use which are sealed to prevent water getting in. Please do not be tempted to use indoor switches outside.

Time Delay Switches

These are common in hotel corridors and communal spaces in apartment blocks, pressing the switch puts the light on, but it remains on only for a short period (most models allow you to pre-set the timer). They are not often used in homes, though the outdoor version can be used to put a light on outside the door while you find the keyhole. Automatic lights operated by Passive InfraRed (PIR) sensors are more popular, despite them being switched on by every passing cat.

Why have boring white plastic switches? DIY stores have lots of sleek, stylish designs to suit any decor.

De-rating for Fluorescents

When a fluorescent tube is first switched on there is a brief current surge many times higher than the normal operating current. This could cause damage to the switch and it used to be common practice for light switches to have two maximum current values, one for conventional loads and a lower one for fluorescents, a system known as de-rating. This is rare these days as switches are better made and able to handle brief surges. Do remember to check the switch specification before you buy, especially on dimmers.

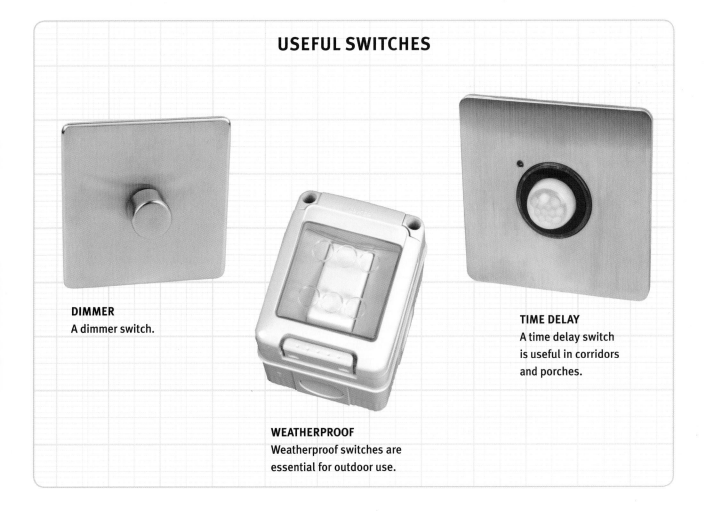

USEFUL SWITCHES

DIMMER
A dimmer switch.

WEATHERPROOF
Weatherproof switches are essential for outdoor use.

TIME DELAY
A time delay switch is useful in corridors and porches.

LOW ENERGY FITTINGS AND LAMPS

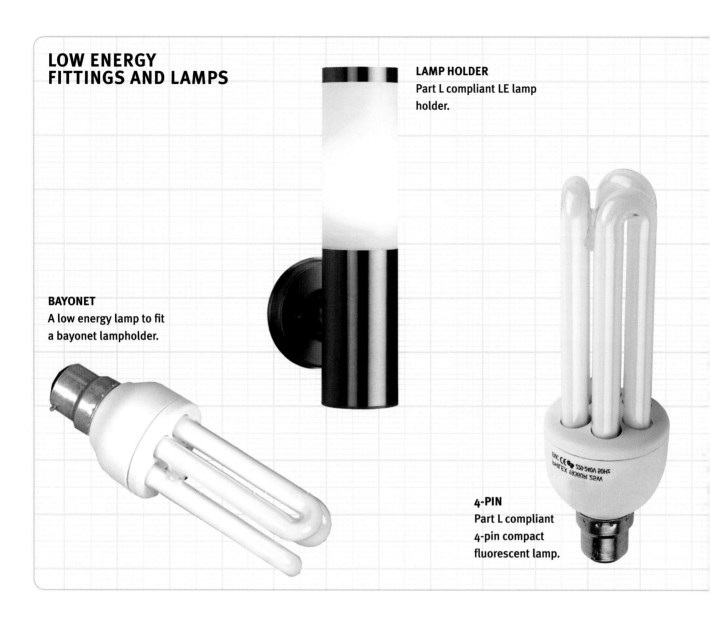

LAMP HOLDER
Part L compliant LE lamp holder.

BAYONET
A low energy lamp to fit a bayonet lampholder.

4-PIN
Part L compliant 4-pin compact fluorescent lamp.

Low Energy Fittings

Low energy (LE) lamps have been around for a while and most are designed to be a straightforward replacement for conventional lamps. They don't actually need special fittings at all, but in an effort to get more people to use them part L of the new Building Regulations (England and Wales) requires builders to fit a certain number of special low energy fittings in each new building (or renovation project). How many depends on the number of rooms, but some builders have decided it is easier, and a good green advertising ploy, to fit low energy fittings throughout. Unfortunately, no one had the sense to define a standard so different

manufacturers produce different fittings and the lamps to suit. Generally they are not interchangeable.

Low energy lamps are actually compact fluorescent tubes which require some electronic circuitry to get them started. Lamps designed to fit standard holders have this circuitry built-in, which is one reason why they are larger.

Some manufacturers, such as MK, produce LE lamp holders which incorporate the electronics so the lamp doesn't have to. Unfortunately, this makes the lamp holder larger and means you cannot use it with a normal UK lampshade. The lamps have pin connections and simply push in

AMP HOLDER
Part L compliant BC3 fittings are the same size as normal bayonet amp holders.

3-PIN
BC3 lamps are similar to normal LE lamps, except for the three pins on the cap.

to the holder. MK lamps have four pins. Some other brands have only two. MEM (now known as Eaton Electric) produce the BC3 (Bayonet Cap, 3 pin) range. In this type the lamp holder is the same size as a conventional BC holder and differs only by having three pins on the bayonet, the electronics is contained in the lamp itself. It could be argued that this means throwing away a functioning circuit when the tube fails, and vice versa, but set against the extra work involved in diagnosing the problem and the complexity of replacing a lamp holder compared to changing the lamp itself, it's the simplest solution. You can use your old lampshades too.

Low Energy Lamps

Low energy lamps for normal lamp holders are easy to find, but as Part L fittings are still uncommon the major retailers will not stock the lamps for them. Both 4 Pin and BC3 lamps can be obtained from Ethical Products Direct:
http://www.ethicalproductsdirect.com
or: www.ethicalproductsdirect.com
or phone: 01740 629940

If obtaining Part L lamps is a problem, or you have the larger MK style holders and want to change them for something slimmer, see the Pendant Lamp Holder project on page 64.

FIT SWITCHES AND SOCKETS

Partition Walls

There are two options, screw a steel box to the studding, or fit a dry-lining plastic box into the plasterboard. In either case, careful marking out and chain drilling will help you make a neat hole.

Brick and Block Walls

Chopping out a space in a plastered wall for a switch or socket box is similar to chasing for cable runs. Switch boxes will generally fit in the depth of the plaster, but socket boxes often need to be cut into the brick or block work. Chain drilling with a hammer drill before chiselling it out is a big help. Fix the box in place with a Rawlplug and screw.

toptip*

If your drill doesn't have a depth stop, wrap masking tape round the drill bit at the appropriate depth, then drill until the tape reaches the surface.

toptip*

In older houses it can be hard to find solid brickwork to take a Rawlplug, so fix the box in place with either No More Nails-type adhesive or Adhesive Plaster as used for dab fixing plasterboard.

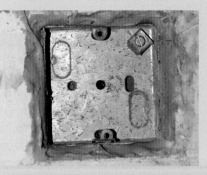

Cables in Walls

Banging a nail into a live cable has been a common cartoon image for decades, but it's not really that funny. To prevent it happening there are regulations about where cables can be placed in walls under floors and above ceilings, and how they should be protected. For more on the law and you, see pages 117–119.

❋ Safe Zones

Unless you are using armoured cable or cable enclosed in steel conduit, which would be unusual in a house, cables should be either buried 50mm deep in the wall or installed in a safe zone. These zones are defined as:

✔ Within 150mm of the top of a wall or partition.

✔ Within 150mm of an angle formed by two walls or partitions.

✔ Horizontally or vertically in line with a socket, accessory or switch.

✔ If you are installing new cables, you must abide by these rules. If you are working in a new house it's reasonable to assume that it was wired correctly, but in older properties it's not unusual to find cables running diagonally under a thin skim of plaster. A wall scanner is a good investment!

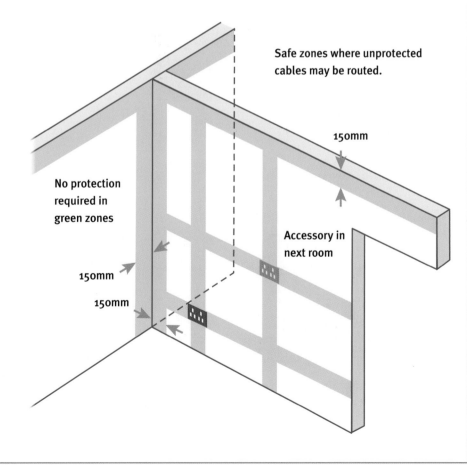

Safe zones where unprotected cables may be routed.

No protection required in green zones

Accessory in next room

150mm

150mm

150mm

If you discover any badly placed cables don't ignore them, replace them. If you are doing other work in the room anyway it won't take much extra time and it could be a life-saver. Cables which are buried without mechanical protection must now be RCD-protected.

doit CHASING

Cutting a groove in a plastered wall to bury cables is called chasing. Dry-lined walls are usually treated the same way but sometimes, especially where old irregular walls have been drylined, there might be sufficient space between the plasterboard to slide a cable behind the board reducing the amount of chasing needed. Partition walls are a different matter, see page 52.

1 Mark out on the wall where the cable is to be buried, remembering to put it in a safe zone. For a single cable, two parallel lines about 20mm apart is ideal, wider if you have more cables **A**.

2 Check the area with a wall scanner **B**.

3 It's easier to cut straight and avoid hacking off too much plaster if you chain-drill the chase first using a hammer drill. You can usually tell when the drill hits the brick. If in doubt, use a depth stop on the drill, or a piece of tape wrapped round the bit to judge the depth **C**.

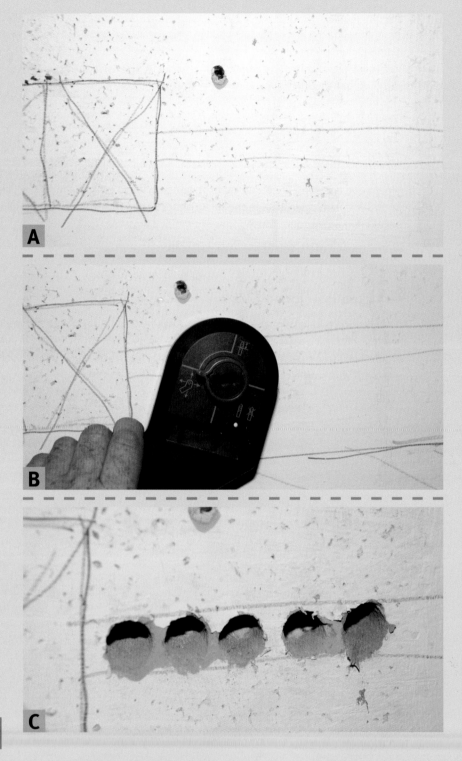

A

B

C

toptip*

If the wall scanner reveals buried cables, dispense with the drill and cut the chase carefully with the bolster. A bolster is not sharp enough to cut through a PVC covered cable without a huge amount of force.

4 Remove the plaster using a bolster **D**.

5 Fix the cable in place with cable clips **E**.

6 Make good the plasterwork **F**.

toptips*

If you are running cables to an existing switch or socket position you can often see where the previous chase was made good, remove the filler carefully with a bolster.

Be warned: NO DRILLING! In a new house, where the cables were fitted before plastering, check the direction the cable leaves the box and carefully chase along it.

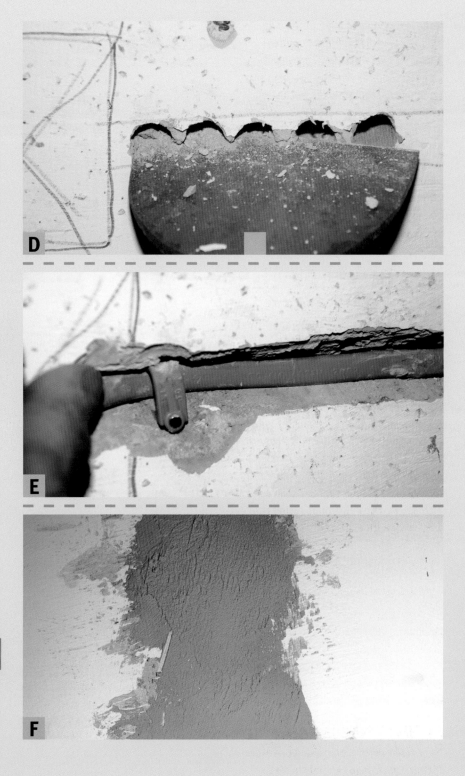

do it   FIT CABLES IN PARTITIONS

Partition walls are commonly made of wooden studding and plasterboard and are usually hollow. Normally you will find there are verticals every 600mm (plasterboard is usually 4in (10.2cm) wide). Horizontal noggins are less regularly spaced, but you can usually find them with a bit of careful tapping. Vertical cables can often be fitted without chasing out the whole length: Mark out the path for the cable as before, avoiding running along one of the vertical studs.

1 Start by carefully cutting away the plasterboard over each of the horizontals, and a little beyond it above and below. If the plasterboard is very thin you will have to cut some of the noggin away to make the chase deep enough **A**.

2 Thread the cable down from above, making sure you go in and out over each noggin **B**.

3 Secure with cable clips on the noggins **C**.

4 Make good the plasterwork **D**.

Horizontal cables are slightly less straightforward but worth the effort to avoid weakening the plasterboard (and a lot of plastering). If you have trouble threading the cable through the holes, try threading piece of stiff earth wire or an unfolded wire clothes hanger through, then pulling the cable through with that.

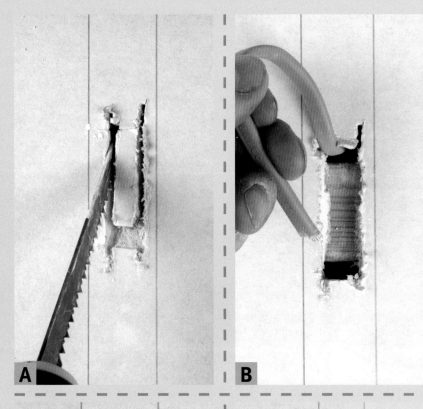

A

B

C

D

INSTALL CABLES UNDER FLOORS AND ABOVE CEILINGS

The space between a ceiling and the floor above is very convenient for electricians (and plumbers and heating engineers) but there are safety requirements here too. Cables should be either 50mm from the surface (above or below) or protected from possible damage.

Where cables run parallel to the joists it is simple enough to fix them to the joists with cable clips, but crossing joists is more difficult. In new builds the normal method is to drill through the joists and thread the cables through, and this is the best method for later additions to the installation. It's not always convenient though, so the alternative is to cut a slot in the top of each joist to lay the cables in. Each slot **MUST** be covered with a 3mm ($1/8$in) thick steel plate to prevent floorboard nails and other fixings penetrating.

The snag with drilling joists, and the reason so many people slot them instead, is that the space between joists isn't wide enough for a conventional power drill and drill bit. Professionals use right-angle drills which as you might expect have the chuck at 90° to the body. These are rather expensive for DIY use, but a right-angle attachment costing around £10 is a reasonable substitute.

toptip*

If you are drilling or slotting joists that already have slots or holes for other cables or pipework, make the new ones at least 150mm from the old, to avoid weakening the joists.

If you are working in the loft space, cables can simply lie on joists, providing the loft is not used for storage. Once you start flooring it and using it then the normal rules apply. Since almost everyone eventually stores stuff in the loft, you might as well do it properly!

RIGHT-ANGLE
Drilling a joist using a right-angle mains drill.

2

Light, Power and Heat

This section shows you how to carry out wiring jobs safely. Please note, those requiring Part P inspection in England and Wales are marked (P), and those requiring a building warrant in Scotland (W). Where you see (P) and (W), this means it is usually OK to go ahead, but if in doubt consult your local authority.

Safety Tips

✳ Staying Alive

✔ Before you start, switch off the MCB (or remove the fuse) which protects the circuit you are going to work on. They should be labelled but to make absolutely sure, leave something switched on, a light or portable appliance, and make sure it goes off when you switch the MCB. You can buy a lock for your MCB and it's advisable to get one **A**.

✔ For peace of mind, as soon as you remove a switch plate, socket or connection point, check all the terminals with a circuit-test screwdriver **B**.

✔ Make sure the rest of the family know what you are doing, the last thing you want is someone helpfully putting the power back on **C**.

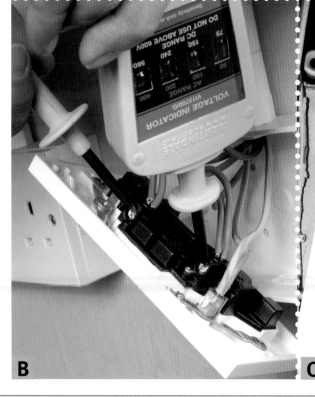

A

B

C

✳ Staying Alive

✔ When you remove switch plates and sockets installed some time ago you often find bare earth wires. Have a stock of green/yellow sleeving handy to cover any you find.

✔ Both wires going to a switch are potentially live, but switches are often wired with standard T&E cable, which prior to 2004 was red/black. At the very least the black should carry a red sleeve (both ends, at the switch and inside the ceiling rose or light fitting). This is becoming rare now though so a better plan would be to use a brown sleeve on both wires.

✔ Remember to run all cables in Safe Zones.

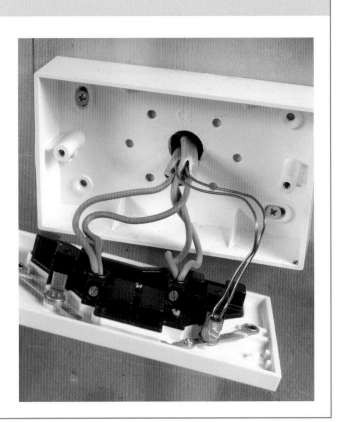

DO I NEED PERMISSION?

P
W

Projects which you are required to inform building control under the Part P regulations (England and Wales) are marked (P).

Those that *might* require a building warrant (Scotland) are marked (W), it usually depends on the location of the work.

Where there is no need to obtain permission or inform anyone, you will see (P) or (W).

In general in England and Wales it is wise to assume that ALL work in kitchens or bathrooms requires building control to be notified. For more information see pages 117–119.

To be absolutely certain where you stand regarding any work, contact your local authority.

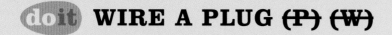

doit WIRE A PLUG ~~(P)~~ ~~(W)~~

Now that most appliances come with moulded plugs permanently attached to their cables wiring plugs by hand is not as common as it was, but there are time when you need to, perhaps because the original was damaged or you want to change the cable, and it's a good place to start your DIY electrical work, so here goes.

1 Open up your new plug **A** with a small screwdriver and check the connections and cable grips. UK plugs always have a cartridge fuse holder attached to the Live terminal. Make sure you have the correct fuse for the appliance in question. It can be easier to wire the plug with the fuse out.

The other terminals connect direct to the pins, but some plugs are better than others at creating a tidy path for the wires. Some plugs use screws to anchor a cable grip, others have stiff plastic jaws that grip the cable. If it's the screw type, remove it.

If you have cut off a moulded plug, or are making up a new cable, you need to strip the cable end. Remove the outer sheath from the cable **B** by cutting carefully with a knife – without damaging the insulation on the individual wires, then pull it off. The amount you need to remove varies in different plugs but it's easier to do too much at this stage and trim the wires back later if they are too long.

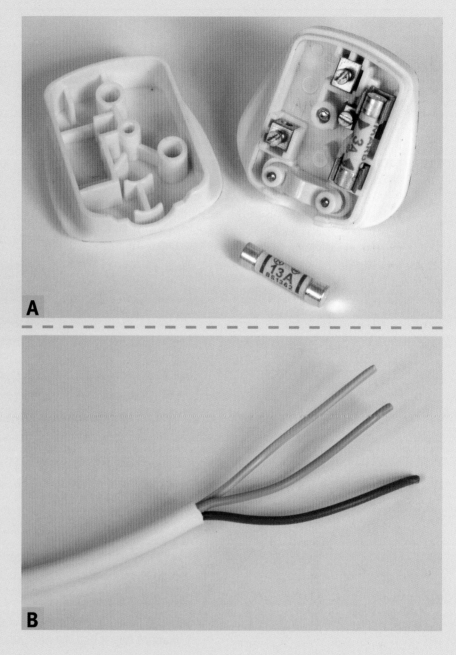

A

B

2 Lay the cable in place and assess how long the wires should be and how much insulation needs removing. **C** In many plugs the Live wire needs to be substantially shorter and the Earth one longer than the Neutral.

If your plug has terminals with holes you only need to bare about 5mm of copper wire for each connection. If you strip more than this it won't all go into the terminal and in extreme cases it can short circuit to other connectors. If you are connecting thin cable, for a small lamp for example it might not sit under the screw very well, in this case strip about 10mm and double it over.

3 Twist the stranded wire in your fingers to make sure every strand enters the terminal, then tighten the screw. Repeat with the other wires. **D**

Some plugs have terminal posts that you wrap the wire round and a nut to screw down on top. These are more expensive but do hold the wire more securely and are good for high current devices. You need to strip about 8mm of insulation, enough to go round the screw once, but not enough to stick out the other side and cause problems.

If your plug has jaws to hold the cable, just push it between them, if it has a screw grip, refit it over the cable. Make sure there is a decent length of outer sheath in the grip otherwise it will slide out and put a strain on the wires.

If you removed the fuse earlier, replace it now, then screw the cover back on.

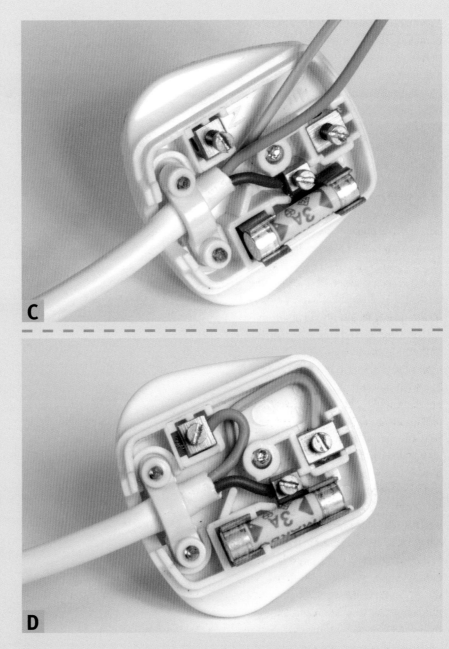

doit WIRE A SOCKET (P) (W)

Wiring fittings such as switches and sockets is a little different, you are dealing with solid cable not flex and there is usually less margin for error since you can't just chop off your mistakes and shorten the cable too often before it no longer reaches the fitting!

1 Thread the cable(s) through the punched out holes into the socket box. If the wall is hollow and you can push spare cable back into it, leave about 300mm spare. If not then leave no more than 100mm **A**.

2 Removing the outer sheath from Twin and Earth cable is easier than flex. Make a small nick on either side of the cable and pull the sheathing apart to expose the Earth wire, then bend it out **B**. Grab it with pliers and pull as shown to split the sheath. Fold back the split sheathing and cut off with a knife or side cutters. Don't go too far, leave the sheath to protect the wires where they enter the box **C**.

3 Use strippers to remove about 8mm of insulation from the Live and Neutral wires. Cut a piece of earth sleeving (yellow/green) and slide it on the Earth wire, leaving approximately 8mm exposed **D**.

4 Insert the wires into the appropriate terminals and tighten the screws. Note that sockets on a ring final circuit will have two cables, maybe three if a spur has been run from the socket. Simply put all the Brown wires into L, Blue into N and all the Earths into E **E**.

5 Finally tuck the wires into the box as neatly as possible and screw the socket to the box **F**.

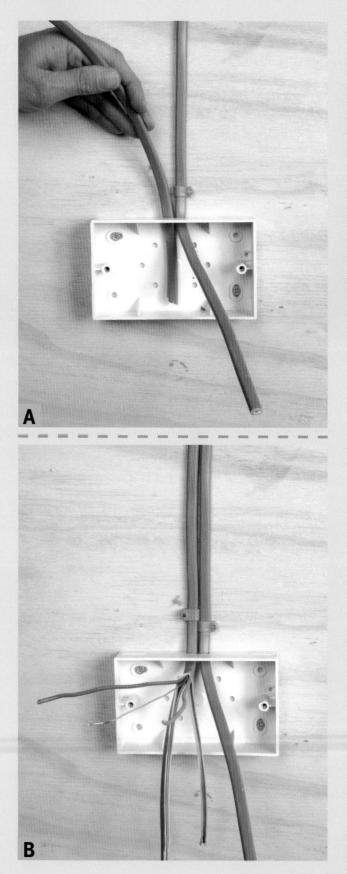

A

B

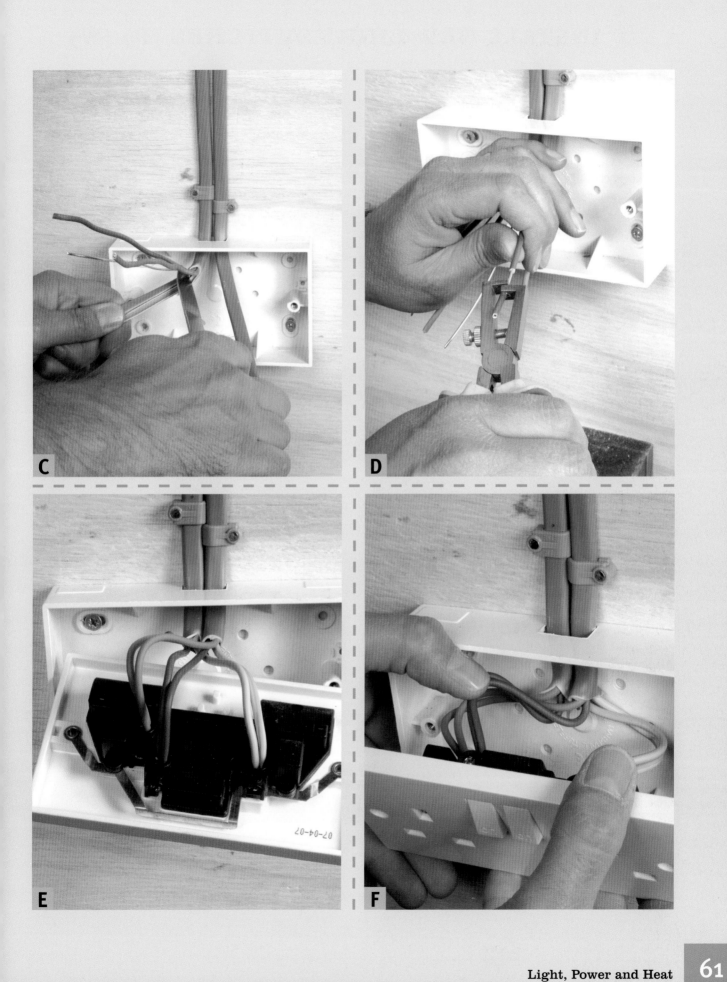

doit INSTALL NEW LIGHT SWITCHES ~~(P)~~ ~~(W)~~

Redecorating a room is the number one reason for making changes to your lighting set-up. There's nothing like new wallpaper or a smart new paint scheme to make your light switches and ceiling roses look very old. A few pounds spent on replacements and a few minutes fitting them can make all the difference. Of course, once you start you might wonder about uplights, downlights, dimmers, low-energy, but we'll start with something simple.

All light switches, recent ones at least, fit onto standard boxes, so replacing like with like really is as simple as unscrewing one, disconnecting it, connecting the new one and screwing it back to the box. Even modern screw-less designs do actually have screws to fix them to the box, they just have a smooth fascia to hide them.

Swapping a simple on/off switch, even a two-way one, for a dimmer is equally simple, but there are a couple of things you should check. First is the total load, most normal light switches are rated at 6A or even 10A, which equate to 1,380W or 2,300W – an awful lot of lights. Dimmers are often 500W, which is still a lot of lights, but it is possible to exceed the rating if you have a large room.

Secondly, if you have fluorescent lamps, including low energy compact fluorescents, you need to make sure they are suitable for use with dimmers. Older fluorescents generally are not and neither are most LE lamps.

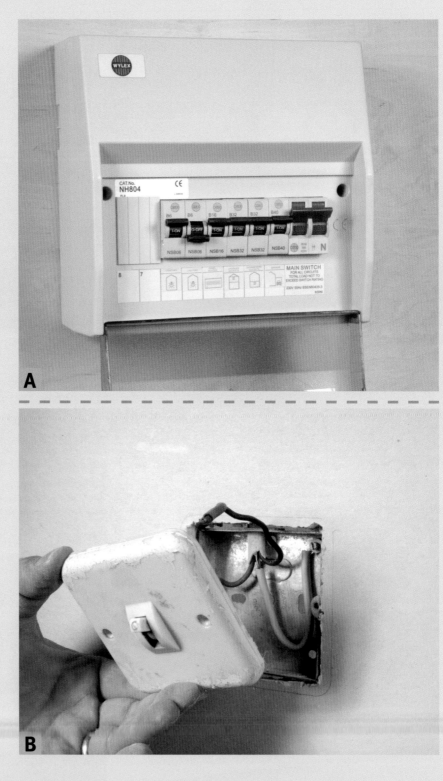

A

B

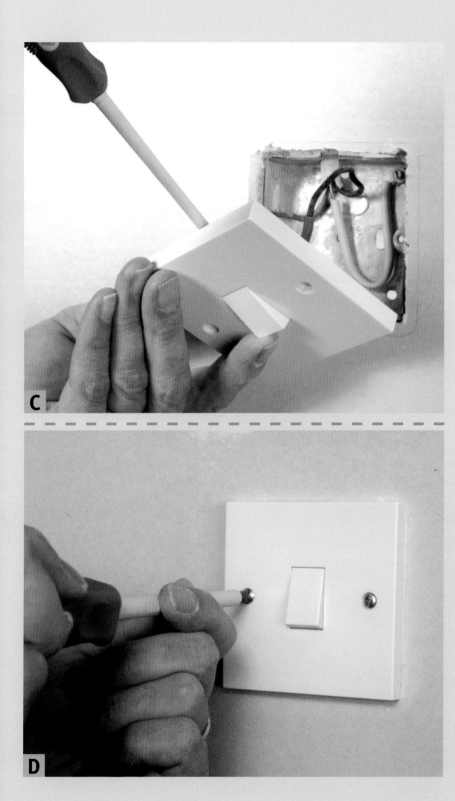

1 Isolate the lighting circuit **A**.

2 Unscrew the switch plate **B**.

3 Disconnect the old switch and connect the new one **C**.

4 Screw the new switch in place and restore the power **D**.

Complications

Ganged switches: Two and three gang light switches can fit in a standard single switch box, but dimmers are larger, two is the limit. The options are either to fit a double box, or an extra single alongside. This is the sort of job that needs doing before you decorate!

Changing from one-way to two-way: Even if one of the switches is going to be where the old one was, you will need to run new cable to it. Once again it's a pre-decorating job. See page 72.

Part L of the building regulations (England and Wales) requires builders to fit some Low Energy light fittings that can only be used with LE lamps. Whilst using LE makes sense for both environmental and financial reasons, many of the fittings are larger than normal and so prevent the use of normal lamp shades. There is no reason why you shouldn't change them for either standard fittings or BC3 LE fittings that are the same size as normal lamp holders. As it seems ordinary lamps will be phased out in a few years and there are plenty of LE lamps for standard holders, the whole Part L thing is rather silly. (See page 46 for more on the regulations.)

1 Isolate the lighting circuit **A**. Remove the lamp and unscrew the top of the pendant lamp holder.

2 Disconnect the flex and remove the lamp holder. Replace with either a standard pendant BC, or a low energy BC3 lamp holder **B**.

Note, you do not generally have to replace the ceiling rose.

toptip*

If you have a stock LE lamp that will be redundant after this modification then it might be worth buying a converter from Ethical Products so you can use 4 pin MK lamps in BC3. They do add to the overall length of the light though.

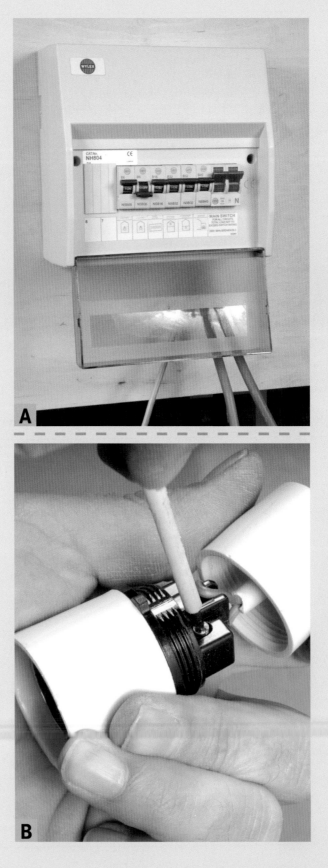

A

B

doit FIT A NEW CEILING ROSE (P) (W)

1 Replacing an old, cracked, paint-splashed rose **A** is almost as simple as replacing a switch. Do make a note of which wire goes where though, there are a lot of wires in a ceiling rose! There should be one or two T&E cables for the radial circuit and a twin live T&E for the switch (or two single and earth if it's a two-way system). You might well find that normal T&E has been used for the switch, and in some cases it's not even sleeved so it's very easy to get them mixed up. Indelible fine-line markers are good for writing on cables.

2 Ceiling roses are not as standardized as switches, each comes with its own unique design of back plate **B**. These have a couple of holes for screws to fix them to the ceiling, but the hole positions can vary. Roses are normally screwed to a joist so the cables can pass down the side of the joist and into the rose via a hole which is offset to one side. If you are lucky, the existing screw holes can be re-used, otherwise you might need filler and a drill.

3 If you are going to the trouble of fitting a new rose, why not buy a ready-wired pendant and rose set? It doesn't take any longer to fit and you get both the style and the safety benefits of a new lamp holder and new flex **C**.

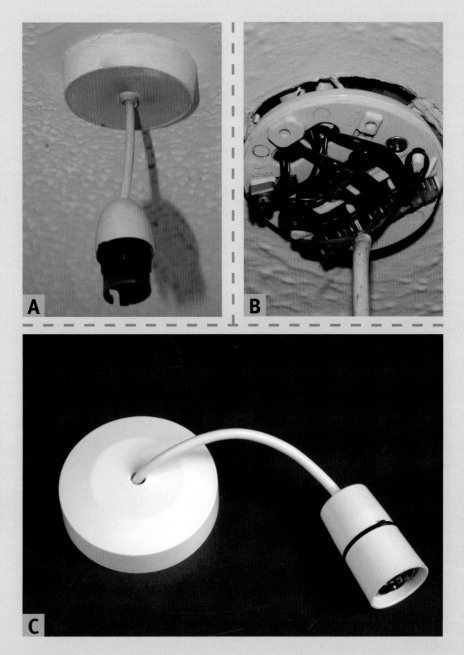

A

B

C

Many lamp fittings, even those that have a boss that looks like a ceiling rose, do not have the necessary connections for a radial circuit and switch connections. Instead, all you get is a short length of plastic terminal block for live and neutral and a screw somewhere on the body for an earth connection. There are a few ways of dealing with these:

1 If the boss is large it may be possible to retain the ceiling rose (or at least its backplate) and fit the new lamp over it **A**.

2 There may be room to fit a longer piece of terminal block which can be wired exactly like the ceiling rose it replaces. You will probably need connectors rated for a higher current, just to make sure the holes are large enough **B**.

3 You might retain the rose and run a loop of flex into the new lamp. This looks OK on chandeliers in grand ballrooms, but not so good close up in a smaller, lower room **C**.

4 Replace the rose with a junction box fitted above the ceiling **D**.

A

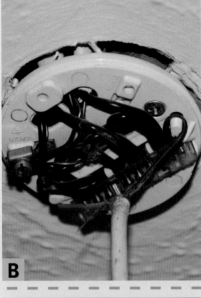

B

C

D

Wall Lights (P̶) (W)

Fitting wall lights is essentially the same as ceiling lights; they don't have roses, so you need a junction box. Junction boxes are normally fitted in the floor or ceiling space and cables then dropped to each light.

You can run several wall lights from a single junction box and switch; if you have a lot of cables a larger 20A power circuit junction box gives you more room.

ENOUGH POWER
A single junction box can serve several wall units.

doit FIT A JUNCTION BOX ~~(P)~~ (W)

Junction boxes are not things of beauty; if you've just bought a smart new lamp you really do not want a junction box stuck on the ceiling next to it.

The most sensible place is in the space between the ceiling and floor above, but access can be a problem. It's bad enough in an older house with traditional floorboards, in newer ones flooring is usually chipboard which comes in large sheets with tongue and groove edges, which makes for a lovely flat floor but is very difficult to lift, especially since it is probably carpeted and has a double bed and a wardrobe standing on it! In some cases partition walls are built on top of chipboard floors, you are never going to get them up.

There are two options: cut a hole in the floor or cut a hole in the ceiling. In either case the hole needs to be large enough to work in, probably around 200mm². Which you decide to do depends on the amount of disruption you can tolerate, and your plastering skills. It is easier to hide rough carpentry under a carpet than to make good a ceiling.

A

B

From Above

1 Some very careful measuring is required to get the hole in the right place. A scanner might reveal where the joists are, but scanning for wood is not 100 per cent reliable. A line of nails is a useful clue, but few are used on chipboard floors **A**.

2 Mark out a square between two joists. Drill a hole in one corner, large enough to accommodate a jigsaw blade. You can use a pad saw, but it's very hard work. You can either cut curves in the other corners, or drill them all and cut in straight lines. Remove the section of floor and do the wiring **B**.

3 To replace the floor, fit a pair of battens as shown. Sawn timber 22x50mm (still commonly known as 'two by one' 30 years after metrication) is ideal. Screw the cut out piece of flooring to the battens **C**.

C

From Below

1 Isolate the circuit and disconnect the existing ceiling rose. The switch cable should be 'twin brown' but if not, and there is a danger the brown sleeving might fall off, label the cable. Remove the ceiling rose 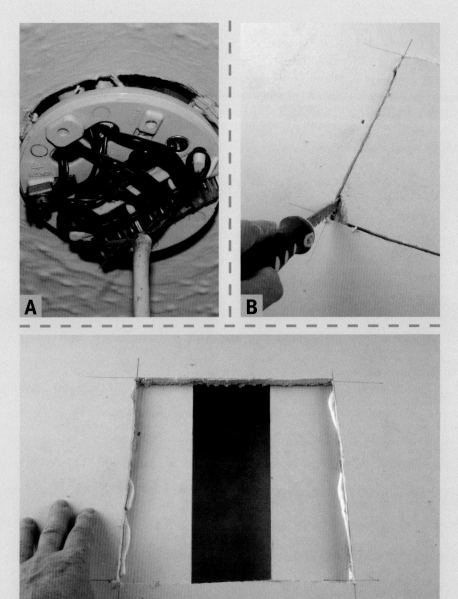**A**.

2 Roses are usually screwed to a joist, so once it's removed you should be able to work out where the joist is. Mark out a hole to the side of the joist and cut it with a pad saw (dry lining saw). The more carefully you cut it, the easier it will be to patch up later. Do the wiring **B**.

3 To replace the cut out, glue two strips of plasterboard or light timber either side of the hole as shown. No More Nails or similar adhesive is ideal. Leave it to dry. Use the same adhesive to stick the cut out back in place, a long batten to prop it in place is useful. Make good with filler **C**.

Use a long batten from the floor to prop the cut out in place whilst the adhesive cures.

WIRE THE JUNCTION BOX

1 If you haven't isolated the circuit already, **DO IT NOW**. Disconnect the existing ceiling rose, label the switch wire if necessary. Remove the rose. Pull or push the cables up into the floor or ceiling space **A**.

2 Open up the junction box and secure the backplate to a joist, halfway between the ceiling and the floor above is the safest place **B**.

3 Cut a length of 1mm² cable to supply the new lamp fitting. Strip the ends. Do not try to cut it short, you can tuck the spare into the floor or ceiling space later **C**.

4 There are four terminals in a junction box. These should be used as follows:
- All the earth wires. Use Green/ Yellow sleeving on them.
- All the neutral wires.
- Live wires from the radial circuit, plus the feed to the switch.
- The return from the switch, plus the live connection to the lamp fitting **D**.

A

B

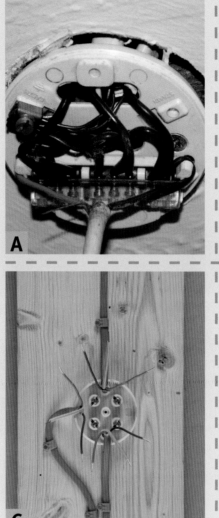

C

D

toptip*

Trim the external cable sheath carefully so no exposed wires are visible outside the junction box (see D). Cut the wires to length to get a neat layout inside the junction box (see D).

5 It doesn't matter which terminal is used for which connection, sometimes the cables seem to lie better one way or another **E**.

6 Secure the cables to the joist with cable clips **F**.

7 Check all the connections, then it the junction box cover **G**.

8 If you are working from above, fit the lamp fitting according to the manufacturer's instructions **H**.

9 If you are working from below, you probably need to make good the hole before finally fitting the lamp, but it would be a good idea to fit it temporarily and test it before re-plastering the ceiling! **I**

10 Put the power back on **J**.

There are various reasons you might want to add two-way switching to an existing light. Being able to operate the light without getting out of bed is useful, especially for elderly or disabled people. Making a new entrance or adding an extension is another – suddenly you are entering the room from another direction and need to put the lights on. If you are not familiar with how two-way switching works, check page 25 first.

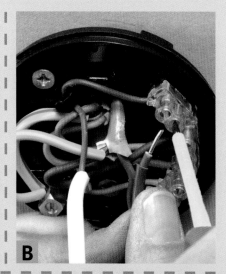

1 First, make sure you isolate the lighting circuit **A**.

2 Open the ceiling rose or junction box, identify the switch cable and disconnect it **B**.

3 Open the existing switch and disconnect the cable **C**.

4 Do not be in a hurry to remove the cable, if the switch is on a partition or dry-lined wall you may be able to use the old cable to pull the new ones through. Try attaching two pieces of wire or twine to the end and pulling it through **D**.

5 If the old cable is fixed or buried in plasterwork, chase it out. **BEWARE**, if there are two or more switches in the vicinity do not use a drill, you risk damaging the cable to the other switches **E**.

6 Fit the new switch box and chase out the cable run to it **F**.

7 Run a Single and Earth cable from each of the switches to the switch terminals in the ceiling rose or junction box (if you were successful in pulling twine through a partition wall, use one to pull this cable). Run a brown/brown T&E cable from one switch to the other **G**. Or use blue/brown and put brown sleeves on both ends of the blue wire. Use other twine to pull this one if possible.

8 Connect them up, single cables to C, T&E to L1 and L2. Fit switch plates **H**. Make good plasterwork.

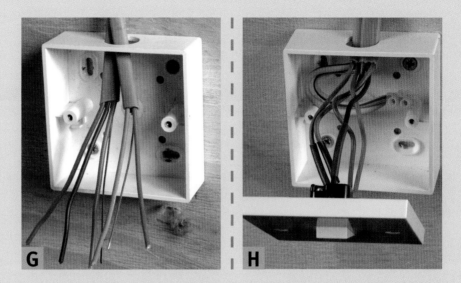

G

H

Anatomy of **Two-Way to Three-Way Switching**

Adding an intermediate switch to a two-way system is actually easier, you just need to break into the twin cable that connects the original two-way switches and insert the intermediate switch. If you want to avoid ripping out the original cables, you can cut the twin cable somewhere convenient and add a junction box.

2   **3**

1

1 Ceiling Rose
2 2 way switch 1
3 2 way switch 2
4 Intermediate switch

2   **3**

1   **4**

FIT NEW SOCKETS (P)(W)

Replacing a socket that is damaged, paint splattered, or just deeply unfashionable is as simple as changing a light switch. Adding new ones is normally quite simple too.

Single to Twin (Easy)

Some builders still fit single sockets to save a few pounds, but with so many electrical gadgets in the average home, doubling them up is a worthwhile exercise. The simplest method is to buy a conversion set. These consist of a shallow plastic pattress suitable for a two-gang socket which is designed to fit onto a flush mounted socket box **A**. They are very simple to install and need no special tools or skill. Fitting one won't mess up your decor and it will solve the problem. They don't look that great though because they are not flush with the wall. They are probably best seen as a temporary solution, to be replaced with a proper flush mounted socket box next time you decorate the room.

A

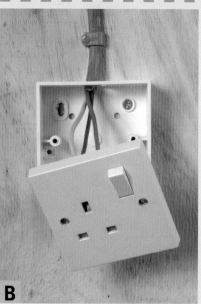

B

If the house is old or you have any reason to doubt the integrity of the wiring, simply opening a socket to check there are two cables isn't proof that it is on a ring final circuit **B**. It might be just the first socket on a dodgy spur with several others attached. See page 116, Trouble Shooting – Spurs and Bridges, for more information.

top tip*

If you use a dry lining box, cut the hole carefully, the flange willl only cover so much.

Single to Twin (Better)

If you work carefully you can often do this without spoiling the decor. In an older house with crumbly plaster it's less likely to be successful, some making good will be required.

1 Isolate the circuit 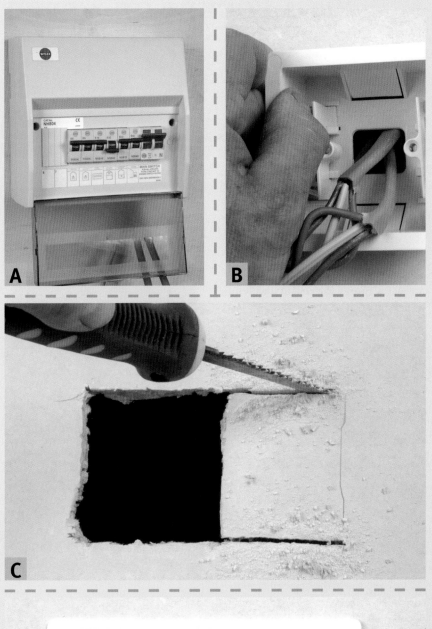A.

2 Remove the existing socket box carefully so as not to damage the surrounding plasterboard B.

3 Enlarge the recess to suit a double size box. A sharp wood chisel will cut the plaster neatly (but need re-grinding after!). Use a dry-lining saw on partition walls. Do this carefully and the socket will cover the cut edge C.

4 Fit new box and socket 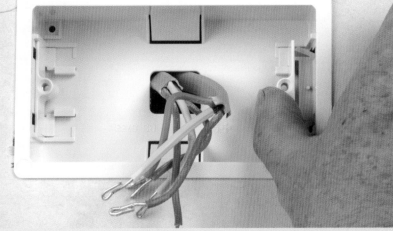D.

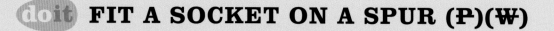

Spurs are generally run from an existing socket, though you can also connect one by breaking into the ring and adding a junction box. Before committing yourself to a particular layout, check the connection you are making really is on the ring, not to an existing spur. Check too that there is not already another spur connected to the same socket. Providing the new socket is not going in an area where there are already cables buried, you can do the preliminary work with the power on. Sweep the wall first with a wall scanner.

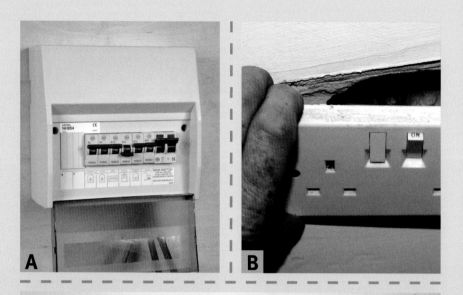

1 Isolate the existing socket by switching the switch on the Consumer Unit **A**.

2 Unscrew the socket from its box and pull it forward **B**.

3 Mark out the position of the new socket and cable run to the closest existing socket **C**.

4 Cut out the recess for the socket box and chase the cable run (take care near the existing socket – hand tools only!) **D**.

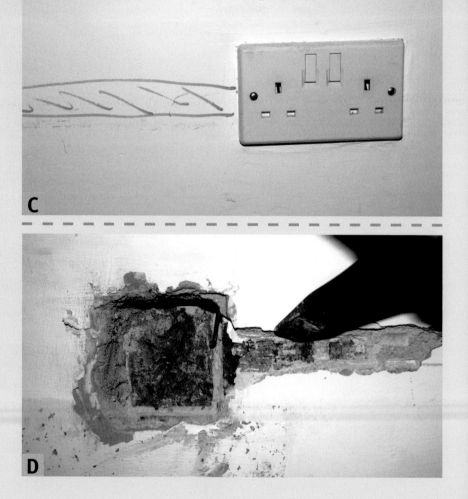

5 Fit the socket box **E**.

6 Run the cable, leaving a few centimetres spare to connect to the existing socket. Fix it with cable clips **F**.

7 Connect the new socket and screw it in place **G**.

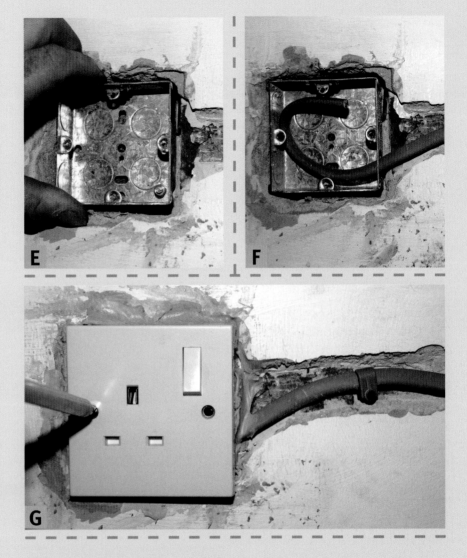

toptip*

Fitting a new socket to an existing circuit does not require notification to building control. It would only require notification if it was in a kitchen, bathroom or outbuilding.

✳ New Sockets Where None Exist

You can add a new socket either by running a spur from the ring main, or by modifying the ring itself. Which you choose depends on several factors:

✔ A spur may only have up to one twin outlet socket, ideal if you just want to run a bedside lamp, not if you intend to power a 3kw electric fire and 42in television.

✔ Installing a spur is easier and makes less mess.

✔ Changing the ring main allows you to fit more sockets.

✔ If you change the ring now, you have the option to add a spur later. You can't run a spur from a spur.

8 Clear any remaining plaster from around the box entry point you are going to use. If necessary, break out a new entry point (a sharp blow from a cold chisel and some wiggling with pliers usually does it) **H**.

9 Feed the new cable into the box, cut to length and prepare the ends **I**.

10 Connect into the back of the existing circuit **J**. Please note: the two pictures 'I' and 'J' have been shot on a plasterboard wall but still demonstrate the connections for adding a socket on a spur.

11 Make good the plasterwork and restore the power **K**.

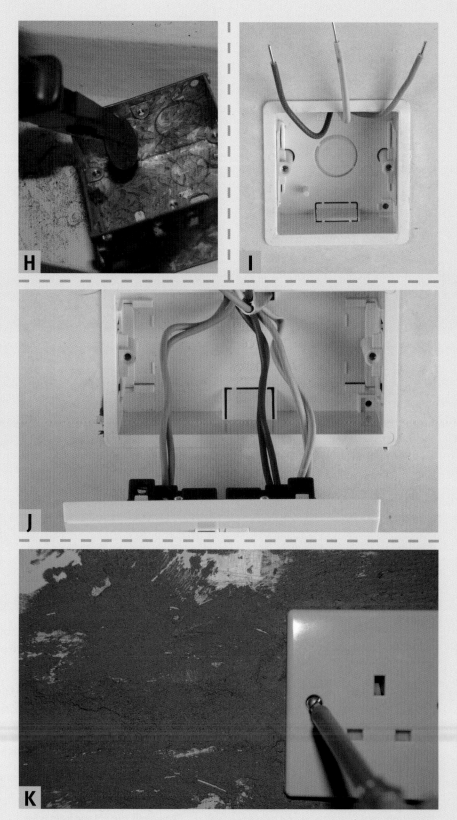

toptip*

Make good plasterwork around the socket box before refitting the socket.

do it ADD A SOCKET TO THE RING (P)(W)

There is no hard and fast limit to the number of sockets on a ring main, although 20 is considered the maximum for good practice. The total load on a ring final circuit should not exceed 30A. Care must be taken to distribute the load evenly around the entire ring final circuit. Several socket outlets in the same position in the ring final circuit may cause overload problems.

Extending a ring isn't difficult, but some planning is needed. Start by opening some sockets or connection points (with the power OFF) in the area where you want to put new sockets, and try to work out the path the cables take. Bearing in mind the safe zones you can usually work out the likely route from the direction the cable leaves the socket box.

If you are very lucky the cable might pass close to where you want the new socket, in which case, follow steps 1 and 2 as above then:

1 Remove one of the existing sockets (probably the closest) and disconnect the cable **A**.

2 Chase along the cable from the original socket until you can re-route it to the new socket position **B**.

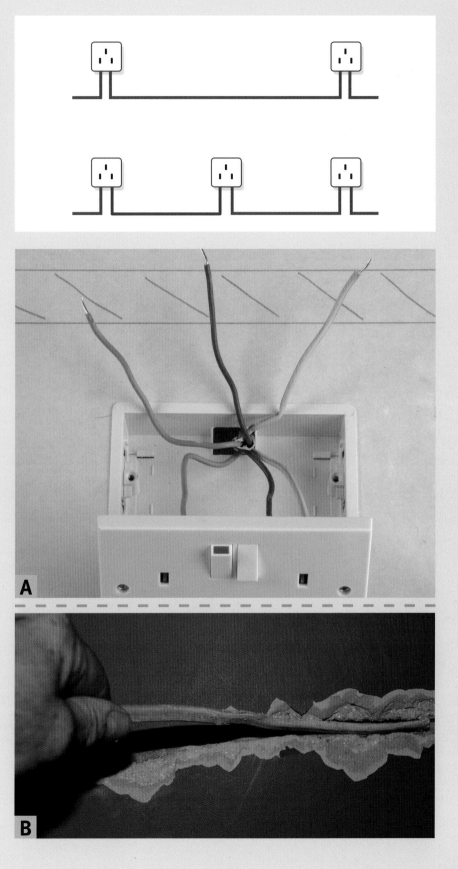

A

B

Light, Power and Heat

79

3 Cut the cable, leaving enough to connect to the new socket 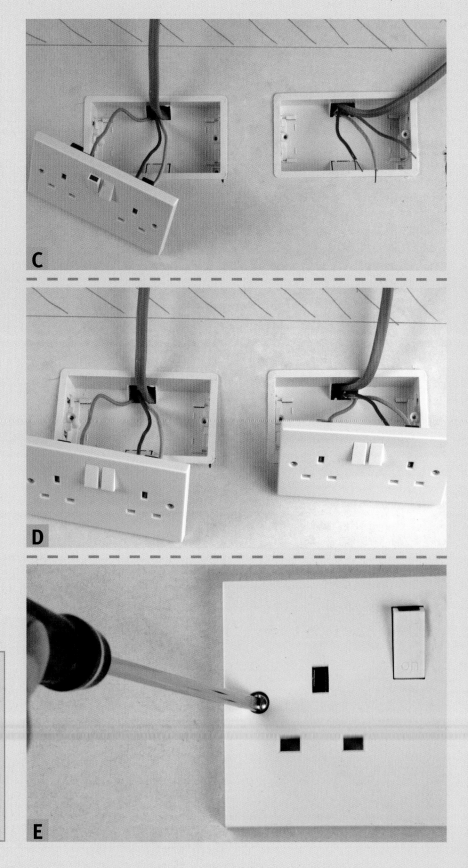C.

4 Run a new piece of cable from the new socket to the old D.

5 Connect everything up and make good E.

WARNING
Do not bunch the sockets at one end of the ring final circuit. This could cause an overload. Make sure sockets are distributed evenly.

Variations

Problem

Cable is close, but not actually passing the new socket.

Solution

You might be able to do as above, providing the distance from the new socket to an old one is less than the distance between the old one and its previous neighbour. It might involve chasing out the cable right back to the next socket though.

Problem

Cable is nowhere near the new socket(s).

Solution

There are three ways of dealing with this:

1 Identify two suitable sockets, remove the connection between them and replace it with two cables leading to your new socket(s). You can leave the old cable buried if it's not in your path for the new ones **A**.

2 Choose one socket, disconnect a cable leading to it and connect it to a junction box, then run cables from the socket and junction box to your new sockets **B**.

3 Fit two junction boxes somewhere in the ring main. This is easy if the cable runs under a floor, but you cannot put junction boxes on a wall and plaster over them **C**!

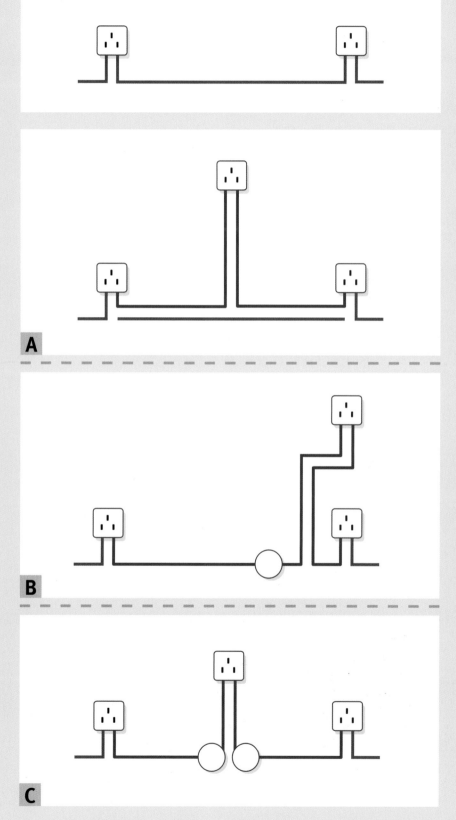

If you have one of the Wylex Standard range of fuseboxes (such as models 204, 304 or 404) it is possible to replace the fuse carriers with modern MCB devices. Wylex produce a range of plug-in MCBs that are the same size as their old fuse carriers and fit directly into the box. Please note, these are not the standard B series MCB for new Consumer Units. Before you start removing the fuse carriers, check the fuse ratings and buy a matching set of MCBs. Other brands of fusebox cannot be upgraded in the same way.

Before working on the Consumer Unit you need to isolate it. It is not sufficient simply to throw the switch on the CU itself. In older installations the CU is connected directly to the meter via short meter tails and the only way to cut the power off is to remove the company fuse before the meter. Your electricity supplier's regulations forbid this and the fuse is fitted with a tamper-proof seal. In this situation you should contact the supplier and ask for advice. Some suppliers will fit an isolator switch for free.

In recent installations an isolator is standard, especially as in modern houses the meter is often located outside some distance from the CU and meter tails longer than two or three meters that cannot be isolated are not allowed by the company's own regulations. If you have an outside meter in a locked cupboard you can get a key from your electricity supplier.

1 Once you've switched off, shut the locker and pocket the key, you do not want anyone else turning it back on! Be warned, this switches off EVERYTHING **A**.

2 Remove the fuse cover and pull out the fuse **B**.

3 Undo the small screw beneath the fuse which secures the shroud and remove it **C**.

4 Replace with the MCB-style shroud with the same colour coding **D**.

5 Insert the MCB (check the colour matches the shroud) **E**.

6 Repeat as necessary. By replacing one at a time you are less likely to get them mixed up **F**.

7 The original cover may not fit back because the MCBs are bigger than the fuse carriers. This is not a safety issue though, plug in fuses are no more dangerous than any other plug or socket **G**.

Fuseboxes like this can be upgraded with Miniature Circuit Breakers.

A

B

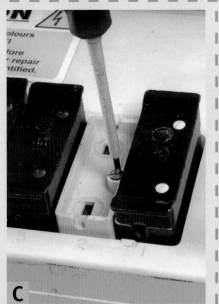

C

D

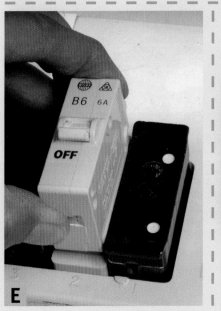

E

F

G

Heating

Most of the electrical work in connection with heating systems is similar to the topics we have already covered. Central heating boilers and immersion heaters each require a radial circuit with an isolating switch and a fused connection point.

Wall-mounted heaters (other than towel rails) are rare these days, but they are occasionally useful in outbuildings where there is no central heating. Wall-mounted heaters should be connected via switched fused connection units.

EASY DOES IT
Underfloor heating is relatively easy to install.

Electrical Underfloor Heating

Electrical underfloor heating is sometimes used in areas where a safe, simple to operate system is required, such as rented properties and holiday lets. If you are refurbishing a property to let, it might be worth considering, and it is relatively easy to do.

The basic system consists of resistance wires which get warm when current passes through them. Calculating the type and length of wire for a given area in a particular building is complex, but generally you buy a kit with the desired heat output. Some kits provide rolls of cable which have to be laid out and taped down, some have the cable fixed to a mesh which makes laying it out easier. Some are designed to be buried in floor screed, others can be tiled over, and some are designed specially for fitting under laminate floors.

The electrical requirements of these heaters are modest. In a small flat you could run a system drawing under 2kw from the ring main via a switched, fused connection point, but generally a radial circuit from a dedicated MCB would be desirable. Connection to the heating system is generally made through a combined connection box or thermostat mounted on the wall, but see the system manufacturer's installation instructions for details.

> **WARNING** BOILERS
> ■ You MUST NOT install a gas boiler unless you are a Corgi registered gas fitter. However, you can prepare the electrical supply prior to the Corgi fitter arriving.

Storage Heaters

Storage heaters require a separate low-tariff meter to make them remotely economical. If they are the only realistic form of heating in your property, get some quotes from specialists. Most of the cost is in the purchase, delivery (they are heavy) and in having a new meter installed. You may find savings you can make by doing some of the work yourself not worth the effort.

If you are interested in doing it yourself, speak to your electricity supplier about their side of the project. They will have to fit an Economy 7 meter, and make the final connection to your system. They won't connect to a system that does not meet building regulations so they will probably charge an inspection fee. In which case it would be reasonable to ask them to provide a Part P certificate, which would mean you don't have to get Building Control approval (England and Wales). Check this beforehand so you know where you stand.

If you decide to go ahead, you will need to provide a second Consumer Unit with ways for each heater fitted with suitable MCBs, then run radial circuits to switched connection points for each heater.

Once everything is installed, arrange for your electricity supplier to fit the Economy 7 meter and connect it to the new Consumer Unit.

Water Heaters

The majority of houses now have central heating and hot water systems, but there are still many that do not, especially small flats, holiday homes and so on. For these places, instantaneous water heaters are ideal. There are basically two types: those with a single outlet which are designed to be positioned over a sink or wash basin, and the multi-point type that are usually fitted under a sink and can supply several conventional taps.

Read the manufacturer's instructions carefully, especially regarding current consumption. Some of these devices are rated at 3kw and can be run from a 13A switched fused connection point on the ring main. But 3kw doesn't heat very much water, and many are more powerful and require a dedicated radial circuit, sometimes as much as 40A. See page 37 on Connection Points or page 93 on Electric Showers for advice on wiring.

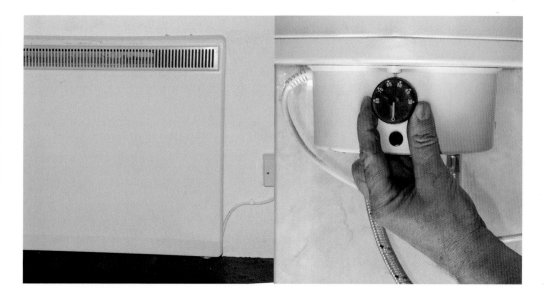

FEEL THE HEAT
(*Left*) Instant hot water is welcome in a holiday home.
(*Far left*) Storage heaters need an Economy 7 meter.

Air Conditioning

Air conditioning is now standard in many cars and workplaces so more people are considering it for the home. British summers seem to be getting hotter, and many systems can also be used to provide heating in the winter. The principle is the same as a refrigerator: a refrigerant gas is compressed until it liquifies, which raises its temperature. It is pumped through a radiator to cool it, then allowed to expand to a gas again, drawing in heat in the process.

In a fridge, the cool end of the process is inside the fridge and the hot end is round the back. In air conditioning, the cool end cools your house and the hot end has to pump heat outside the building. If the system is reversible it can extract heat from outdoors even on a cold day and pump it inside. There are several ways of achieving this. A common single room system has two separate units, one inside and one outside, joined by pressurized hoses. These are very neat but require careful fitting, both halves

AIR CON UNITS

A variety of different air conditioning units are available.

KEEP COOL
Built-in air conditioning
provides unobtrusive
cooling and heating.

of the system are pre-pressurized and the hoses have snap-on connectors which connect with minimal loss of pressure, but you only get one chance to get it right. Systems like this use around 6.5A and can be connected to a switched fused connection point on a ring main. Easier to fit are portable single unit systems. These have both halves of the system in a single box with the warm air being blown out of a flexible hose similar to those used by tumble driers. They can be run from a 13A socket and some have adaptors to fit the hose into a window. They are useful temporary solutions, but not exactly attractive. With some ingenuity you could probably disguise one inside a cupboard.

More complex systems can be fitted in the loft space of a house. In the system shown here, warm air is drawn in from the stairwell, cooled, dehumidified and fed back into the house, the waste heat is blown out through the end wall. These are more expensive systems to install but all the machinery is hidden away in the loft. Being larger than a single room unit you will need to run a radial circuit from the Consumer Unit to an isolating switch and then to the air-conditioning unit.

3 Indoors and Out

Water is a surprisingly good conductor of electricity so you have to take special precautions when wiring in kitchens, bathrooms or outdoors. It's not particularly difficult, just make sure you use the right fittings, and follow the tutorials carefully.

Bathrooms

Any electrical work in a bathroom other than replacing an existing light fitting should be referred to building control (England and Wales).

There is an old saying, 'Electricity and water don't mix'. Actually they mix only too well, water is quite a good conductor and the combination of wet hands, electrical fittings and well earthed pipework can be fatal. Bear the following in mind and you won't go far wrong:

All switches (lights, electric shower, radiator, extractor fan and so on) should be operated by a pull cord (and the switch body should be out of reach, preferably on the ceiling). Actually the regulations define zones in bathrooms where various things can be fitted or not, but the zones are larger than most bathrooms so in practice our two rules will keep you safe and legal. Shaver points that have a built-in isolation transformer are permitted.

Under the new wiring regulations, a 13A outlet may be fitted in a bathroom. It must be at least 3m horizontally from the edge of the bath and must be RCD protected. All circuits in bathrooms must now be RCD protected.

BAD MIX
Keep electrical fittings away from water.

SPACE
A larger bathroom will
provide more scope for
home wiring.

Anatomy of **Bathroom Zones**

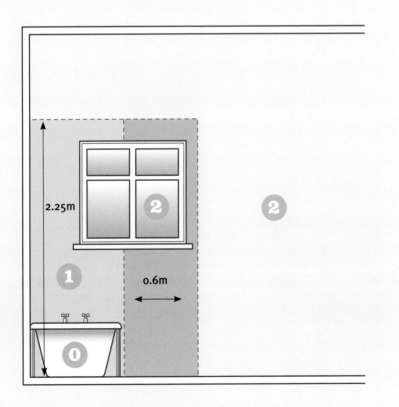

2.25m

0.6m

1

2

2

0

0 Zone 0
The interior of the shower or
bath which can hold water.

1 Zone 1
The area directly above Zone
0 within a limit of 2.25m
above the bottom of the bath
or shower. The zone also
runs 1.2m horizontally from
the centre of a shower outlet
to the height of the outlet or
2.25m, whichever is the
higher.

2 Zone 2
The area beyond zones 0
and 1 (which are 0.6m
horizontally and up to 2.25m
vertically). Zone 2 also
includes any recessed
window with a sill next to
the bath.

do it FIT AN ELECTRIC HEATER, TOWEL RAIL OR RADIATOR

No one likes a cold bathroom so electric heaters, especially towel rails, are popular. They are designed to connect to a fused, switched connection point, but of course you can't have a switch in the bathroom. Here's how you do it.

1 Check the heater manufacturer's instructions about siting it, then work out where to put a flex outlet convenient for the heater. Fit the flex outlet, recessed if necessary, and the heater. Isolate the ring final circuit **A**.

2 Find a sensible location outside the bathroom for the fused, switched connection point, ideally near an existing socket, or at least near to the ring main **B**.

3 Connect the connection point to the ring main. Run a spur from the connection point to the flex outlet. Fit the flex outlet and connect the heater **C**.

Alternatively...

Use a pull cord switch for the heater.

PLEASURE

A warm towel is one of life's little pleasures, and an electric towel rail is simple to fit.

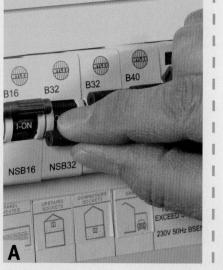

A

B

C

do it FIT AN ELECTRIC SHOWER

Electric showers use a lot of power and cannot be connected to the ring final circuit. They have to have their own radial supply from the consumer unit. If you are lucky there may be space in the unit for you to fit a suitable MCB.

Installing a shower is both a plumbing and an electrical job. If you are not confident in both fields, get a professional in. Plumbing work is outside the scope of this book, but the wiring work isn't complex.

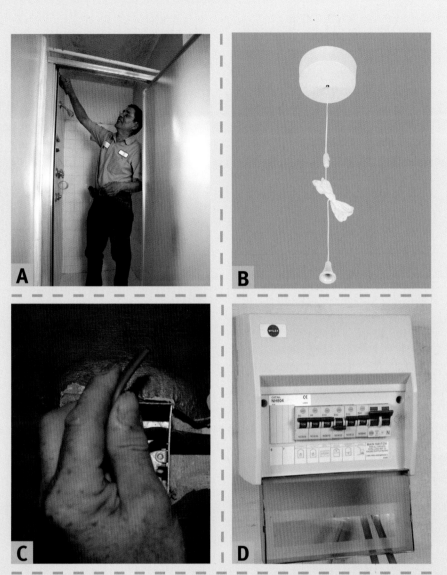

1 Read the manufacturer's installation instructions before starting work. In particular, check the current rating of the shower and the recommended cable thickness. Fit the shower and connect the pipework. Make sure the shower unit and pipework are bonded **A**.

2 Fit a suitable double pole shower control switch. These are normally pull cord operated and fixed to the ceiling. In rare cases you might use a conventional isolating switch, fitted OUTSIDE the bathroom **B**.

3 Chase out a cable run from the Consumer Unit to the switch and from switch to shower. If you start the wiring from the shower end you can do most of it with the power on **C**.

4 Switch off the main switch on the Consumer Unit before connecting the cable **D**.

5 Check the manufacturer's instructions BEFORE trying the unit, you may need to flush water through it before turning on the power. Switch on and enjoy! **E**

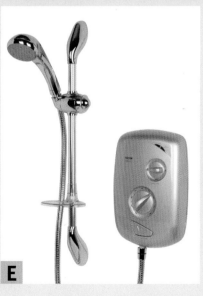

FIT A SHAVER SOCKET

Shaver sockets have the same size fascia as a twin socket but the socket box has to be 45mm deep. It also has to be mounted vertically instead of horizontally. The boxes are available in the usual forms but, given the depth required, a surface mounted box looks rather hefty. Another option is to buy a bathroom cabinet with a shaver socket already built-in, some of these have lights too, which saves you wiring them separately.

Because shaver sockets are limited to 10VA (see page 38) you can connect them into a radial lighting circuit, which is useful as there are seldom any ring mains in the bathroom. In the example here, we are assuming you have a bathroom upstairs with access to the lighting circuit in the loft.

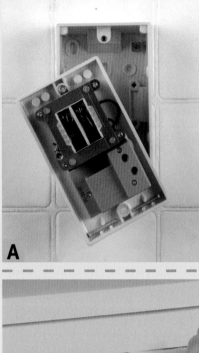

A

B

1 Mark out where you want the socket and fit the box **A**.

2 If the box is recessed into the wall, cut a chase up to the ceiling for the cable. If you are using a surface mounting box, you might consider using plastic trunking to run the cable on the surface too **B**. Make a hole in the ceiling for the cable.

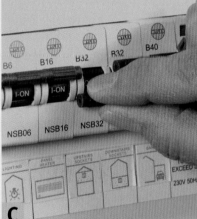

C

D

3 Now isolate the lighting circuit **C**.

4 If you have a ceiling rose in the bathroom, open it. Assuming it is wired conventionally you can connect 1mm² cable to the radial circuit terminals, pass it up into the loft, route the cable across to the wall and down to the socket **D**.

5 If you don't have a ceiling rose, or the rose is just a slave fed from a junction box, then you need to locate that and connect there. In some cases you may need to add a junction box **E**.

6 If you are connecting a cabinet with built-in lights and shaver socket, consider connecting to the switched side of the rose so the lights go on or off with the main switch. The downside is that you can't shave without turning the lights on.

7 Fit the socket and make good any plasterwork.

E

SAFETY

Only approved low-power shaver sockets should be used in bathrooms.

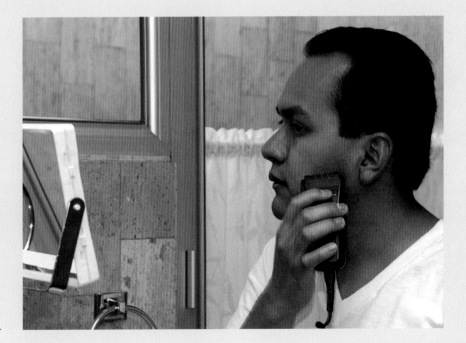

FIT AN EXTRACTOR

There are two basic forms of extractor for bathrooms, those designed to fit into a wall or window and run from a fused connection point, and those combined with a light designed to fit into a ceiling, usually above a shower.

Wall and Window Extractors

Like wall mounted heaters these are connected by a short length of flex to a flex outlet. Many extractors have a pull switch built-in, so they can be connected to an un-switched fused connection point. If not then you must use a pull switch.

Ceiling Extractors

Most of these are designed to be safe in the steamy environment of a shower enclosure, but do check before you fit it, some are not.

Positioning the extractor is important. The regulations define zones, based on proximity to the bath or shower, and a list of which equipment may or may not be installed in each zone. Zone o is the area occupied by a shower tray or interior of a bath, where nothing can be installed. Above it is Zone 1, where you can install showers which are designed to prevent water getting in, but not fans or light fittings. However, the ceiling above the shower (providing it is higher that 2.25m) is Zone 2, and you can install a light fitting or extractor fan there. Follow the manufacturer's fitting instructions carefully.

A small fan draws less current than a light bulb, so it is safe and legal to connect them to a lighting circuit. This is useful as most bathrooms are upstairs and there is rarely a ring main in the loft, but there will be a lighting radial circuit. It also means you can connect the fan to the ceiling rose or junction box so that it comes on when you switch on the light.

Some extractor systems are more sophisticated and include a timer that runs for a fixed period of time, then switches off automatically. These need to be connected to the live side of the ceiling rose or junction box so they can run after the light has been switched off. These extractors should be connected via an isolating switch, normally a pull switch.

A three pole isolator is required to isolate the fan if it is timed from the lighting circuit. This must isolate the line and neutral feed to the fan but allow the bathroom light continuation of service.

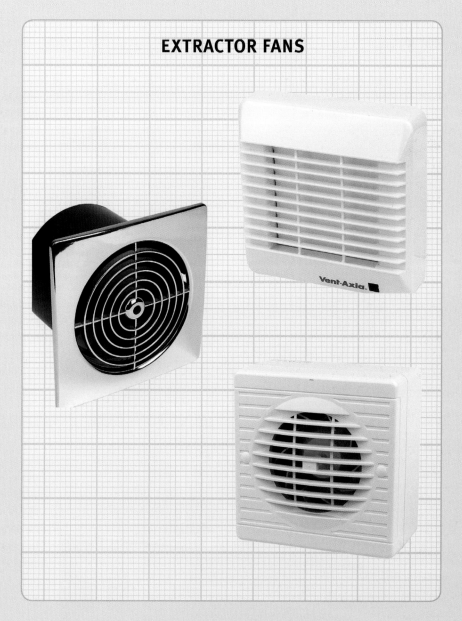

EXTRACTOR FANS

Kitchens (P) (W)

The number of electrical gadgets in the average kitchen increases every year and when you find yourself unable to have a coffee because the breadmaker is on, it's time to add some more sockets. Adding sockets has already been covered, but there are particular issues in kitchens. Although less dangerous than bathrooms, there is still that combination of electricity and water to worry about.

To avoid water splashing into sockets and switches, a minimum distance of 300mm (horizontal) from sink to fitting is recommended, 600mm is even better. Keeping sockets and switches to a minimum of 300mm from cooker hobs is also considered good practice. Sockets and switches over work surfaces should be a minimum of 150mm above the surface to reduce the likelihood of them being affected by splashes and spills.

It is obviously easier and neater to have sockets for large appliances such as fridges, washing machines and the like hidden under the work surface or even behind the appliance. This is permissible, but you must be able to switch the appliance off without having to drag it out from under the work surface. ANY work undertaken must be reported to building control and the required certification by a competent person must be submitted.

toptip*

Kitchen extractors can be built into a high level kitchen unit, fitted between two units or completely independent. Extractors connect via a short length of flex to a fused, switched connection point. Some manufacturers recommend using an isolation (double pole switch). Always read the manufacturer's installation instructions thoroughly.

CAUTION ELECTRIC SAFETY IN THE KITCHEN

■ Any alterations to the kitchen circuits must comply with the latest regulations.

■ If there's an electric point, such as a socket or a switch, within 800mm of the tap, it must be moved.

■ Electric points should be at least 150mm horizontally away from a hob.

■ Plug sockets should be a minimum of 450mm and a maximum of 1,200mm from the floor. The centre of a plug socket must be at least 150mm above a worktop.

■ Surface mounted plug sockets can be installed high up in base units where they will not be obstructed by the cabinet contents. They can be incorporated into an extension of the ring main or as a spur.

■ If the supply to an appliance cannot be installed to the wall inside an adjacent cabinet, consider extending its cable.

■ The alternative to a plug socket is an FCU (Fused Connection Unit). These incorporate a cartridge fuse, and can have an on/off switch. From the outlet, a cable can be run to the wall behind the appliance and terminated in a cable-outlet plate. They can be installed on the wall above the worktop to control an appliance below.

■ If the connection is made behind an integrated appliance, there must be an additional isolation point in an accessible position. Use an FCU.

■ Consider a cabinet light system as another appliance. Power can be provided from a plug socket or an FCU.

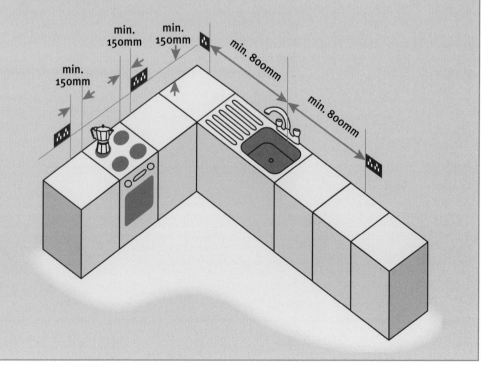

For general advice about adding extra sockets, see page 82. You have several choices when it comes to fitting sockets under work surfaces: fit them to a wall alongside the appliance in a position where they can be reached safely; fit them to a wall inside an adjacent cupboard (socket outlets must NOT be fitted to kitchen furniture); or use unswitched sockets out of sight, connected via switches.

The second choice requires a hole about 60mm diameter in the side of the cupboard for the plug to pass through. The third option is neater and quite simple, but best tackled at the same time as a major kitchen upgrade.

Many builders and kitchen fitters use isolation (double pole) switches in this situation and you often see them engraved with 'Fridge,' 'Washing Machine' and so on. They are usually white and often have red switch rockers. They are marginally safer than conventional connection points but not pretty and not obligatory. If you prefer to use switched connection points that match the rest of your fittings that is up to you. Most engraving shops will be able to engrave them for you.

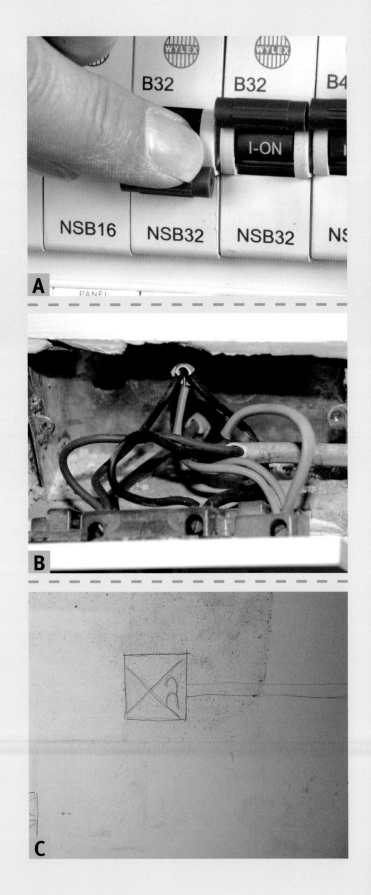

1 Isolate the kitchen socket circuit **A**.

2 Open some existing sockets to discover how the ring is laid out, remember you cannot add extra sockets to spurs **B**.

3 Work out where to put the switches and sockets. You might prefer the switches on the wall directly above the relevant appliance, or you can group them all together **C**.

4 Chase out the plaster as necessary and fit the socket and switch boxes (although they are 'switches' connection points need deep 'socket' boxes) **D**.

5 Connect the switched connection points into the ring main and run a spur from each to the relevant socket **E**.

6 Make good the plasterwork and restore the power **F**.

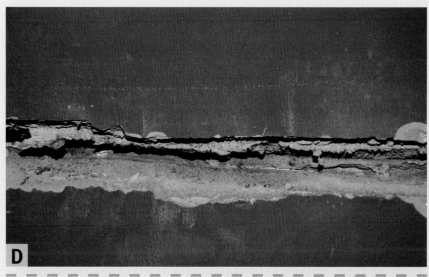

D

E

F

toptip*

If you are fitting hidden sockets for a washing machine or dishwasher DO NOT fit them under or close to a sink where they might get wet

 ADD A FREEZER SUPPLY

1 One unfortunate effect of having sensitive RCD and MCB devices in modern Consumer Units is that a minor fault or power surge can trip the circuit breaker and destroy hundreds of pounds worth of frozen food. Some houses already have split load consumer units with one side of the system supplied via a conventional isolator switch rather than an RCD reserved for the freezer **A**. If you don't have this luxury, why not install a separate freezer supply?

2 The starting point is a small consumer unit such as the Wylex NH206/63 with an isolation switch rather than an RCD. This is a two-way device into which you can fit one or two Type B MCBs. (An example of a two-way consumer unit which may be added to extend the installation can be seen below **B**.) A freezer is unlikely to draw more than about 2A for any length of time, though when the motor first starts it will draw a lot more for a brief period. A 6A MCB is likely to be sufficient for most freezers, but do check the freezer manufacturer's specification which is usually hidden on a little plate at the back.

The cable from the new CU to the socket could be as light as 1mm², but it would be good practice to use much heavier into the CU itself, giving you the option to use the other way for something else in future.

3 With light cable and a low current MCB device, it would make sense to mark the socket 'Freezer Only' **C**.

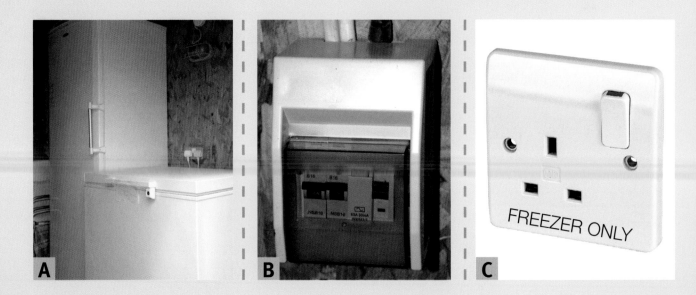

A   B   C

ADD A COOKER CONTROL UNIT

A cooker control unit is just a large switched connection point run on a radial circuit direct from the consumer unit, where it has its own MCB. Between the control and the cooker itself is a giant-sized flex outlet known as a terminal outlet box or cooker connection unit. If you are redesigning your kitchen layout you may well have to move the outlet, the switch or both.

When planning the kitchen layout, connection units are normally fitted behind the cooker, and switches close to the hob, but not over it. About 300mm away is reasonable, close enough to reach easily in a crisis but not so close you will burn yourself reaching across the hob. Like any switch or socket it should be 150mm above the work surface.

Cookers use a lot of current, especially if both hob and oven are electric, so check the manufacturer's installation instructions for guidance on cables and MCB ratings. 6mm² cable and 45A MCBs are common requirements.

toptip*

If the wall scanner reveals buried cables, dispense with the drill and cut the chase carefully with the bolster. A bolster is not sharp enough to cut through a PVC covered cable without a huge amount of force.

BE SAFE
Keep your sockets and switches a minimum of 300mm from sinks and a minimum of 300mm from hobs.

toptip*

Old-style cooker control points had a socket for heavy duty appliances when other sockets were scarce and maybe on 2A or 5A. These days they are unnecessary and trailing flex across the cooker is NOT a good idea, so fit a modern control point.

1 Work out the location of the control, connection unit and path for the cables **A**.

2 Chase the cable paths and cut recesses for the control and connection unit **B**.

3 Fit the boxes and run the cable from the connection point to the control and from there to the Consumer Unit, connecting the cables to the cooker control point **C**.

4 Connect the cables to the connection point. If you are ready to fit the cooker, connect that too **D**.

5 Switch off the main switch (RCD) on the Consumer Unit before making the connection **E**.

6 Make good the plasterwork – and finish the rest of the kitchen! **F**.

toptip*

Work backward from the new fitting towards the Consumer Unit, so you can leave the power on until you have to make the final connection.

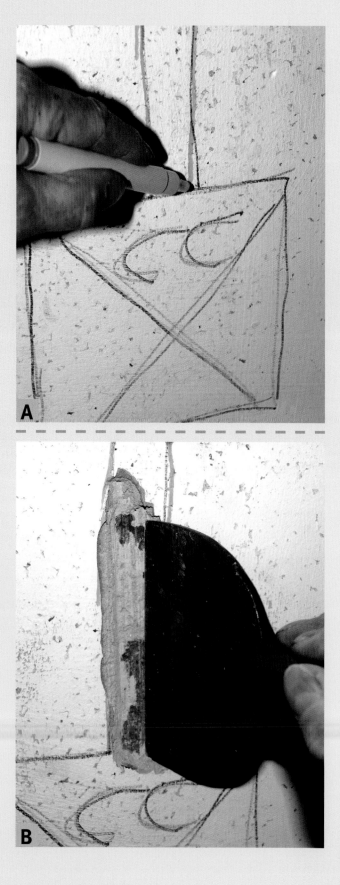

A

B

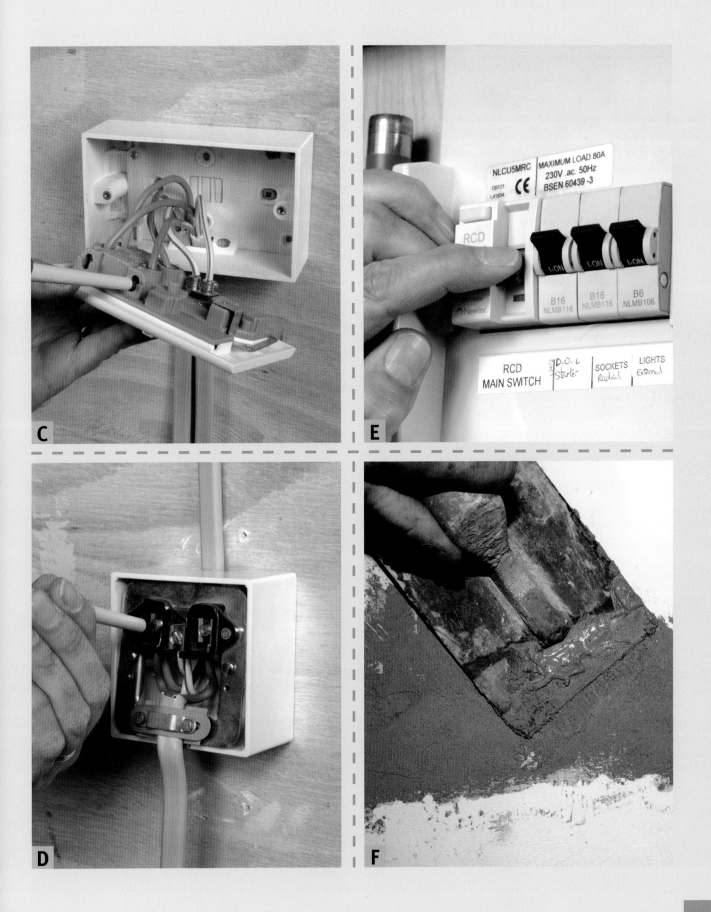

C

D

E

F

Outdoor Power Projects (P) (W)

The use of electricity outside the house is booming. Where once there might have been a porch light now we have floodlit patios and spotlights on specimen plants. Instead of a shed there are garden studios and home offices. Hot tubs and water features have replaced the old pond, and that's before you count all the electric powered mowers and blowers, trimmers and strimmers. An extension lead out of the kitchen window doesn't really cut it any more.

Electricity is dangerous outdoors for the same reason it's dangerous in kitchens and bathrooms – water. Properly designed outdoor fittings are essential, and they must be fitted according to the manufacturer's instructions, to keep water out. The other hazard out there is people: you, your friends and family. Slicing through extension leads with the lawn mower or hedge trimmer, putting a spade through a buried cable, using a power tool in the rain. You can't prevent accidents, but

you can make your electrical installation inherently safe so that accidents don't kill people. The first thing to get right is the choice of cable. You can either use armoured cable (sometimes known as SWA or Steel Wire Armoured cable), or conventional wires inside a steel (not plastic) conduit. Plastic doesn't provide much protection and rapidly becomes brittle when used outdoors.

Cable and conduits shouldn't be attached to temporary structures such as trellises and light fencing, when they fall down in a gale it puts too much strain on the cable. Armoured cable can be buried, but deep enough to stay safe during normal gardening activities. Putting cable inside reinforced hosepipe, as demonstrated by TV gardeners, is only for low voltage supplies. It's not sufficiently strong to stop a determined gardener with a sharp spade.

It's important to note that you will need to notify the authorities about all or any work outside.

SAFETY FIRST
Ensure your electrical work is safe in case of accidents.

 **HANDLE ARMOURED CABLE**

Inside armoured cable is conventional flex, but stripping the steel reinforced outer without damaging the flex is a little tricky. It's a good idea to practise this on the end of the cable before you cut it to length and spoil it!

1 Decide how much sheath you need to remove, bearing in mind that the flex should only be exposed inside the fitting **A**.

2 Use a junior hacksaw to cut through the outer plastic sheath and part way into the steel reinforcing wires. Roll the cable to cut right round it. Don't try to cut through the steel, you just have to notch each wire **B**.

3 Use a knife to remove the outer sheath as far as the cut **C**.

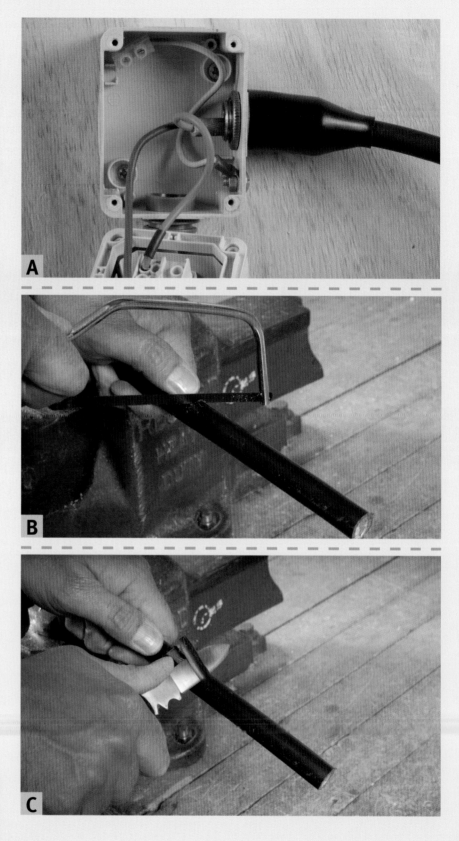

A

B

C

toptip*

When removing the outer sheath, take care to cut away from yourself.

4 Take hold of a few of the steel wires at a time and bend them back and forth until they snap where you notched them **D**.

5 Strip the inner flex like any other **E**.

When connecting cable into outdoor fittings, follow the manufacturer's instructions regarding glands, sealants etc. Make sure the cable is securely fastened to a wall or post, making sure there is no strain put on the gland or fitting.

Do not try to put sharp bends in armoured cable, you cannot do it without damaging it. Look at the size of the roll it came in and don't try to bend it to a tighter radius than that.

D

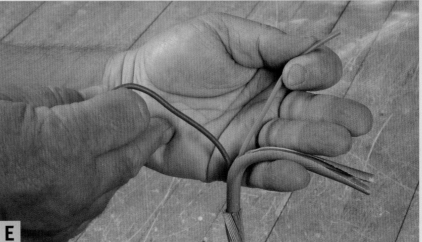
E

Circuit breakers

Any circuits used outside should be protected by an RCD. This can be provided by the main RCD in a modern Consumer Unit, but if you are planning a lot of outdoor circuitry an extra CU in an outbuilding (or weatherproof enclosure) is a good idea. For less ambitious projects you can run cables from an existing CU if you have spare ways or a secondary CU in the house.

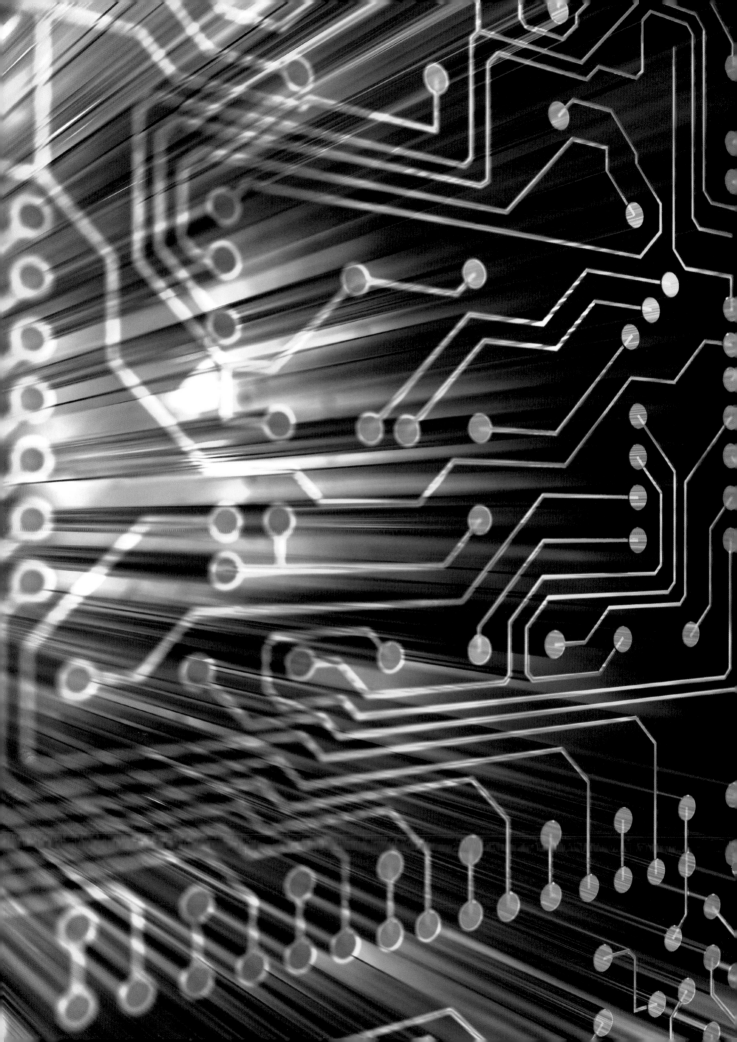

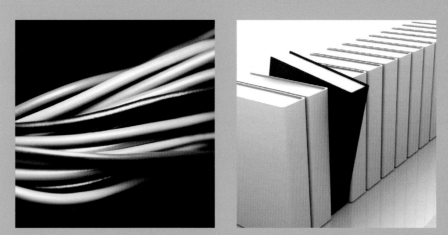

4 Troubleshooting and the Law

Two things worry people when doing their own wiring jobs: (a) it might not work and (b) if it's not done right they may not be able to sell the house when the time comes. This section is designed to put your mind at rest.

Troubleshooting circuits isn't difficult, but does need a logical approach. Most problems are just variations on those described below. For basic troubleshooting, a simple multimeter is ideal.

Dead Sockets

If all the sockets and connection points on a ring are dead it is likely the MCB has tripped. This may be just a freak occurrence, so try resetting it. If it won't reset, switch off and unplug all the appliances and try again. If it resets now, then the problem lies with one of the appliances; plug them back in one at a time until you find out which one.

If the MCB will not reset even with everything unplugged, then the circuit itself is faulty. This is, thankfully, rare but can be a trial to track down. Before doing anything too technical, inspect all the sockets for signs of burning or damage and replace any that look suspect.

If that doesn't resolve the problem, turn the power off and open each socket or connection point and inspect the cabling. Loose or burnt wires are usually obvious, but check for un-insulated earths too. If a bare copper earth wire touches a live connection it will trip the MCB, so make sure they all have sleeving. Any burnt or damaged cable should be replaced.

Test the Live, Neutral and Earth conductors

If you still have a problem then you need a multimeter and some patience. Leave all the appliances unplugged. Turn off the main switch (RCD) and disconnect both ends of the ring from the consumer unit. Set the meter on Resistance and touch both Live connectors, there should be virtually no resistance. If there's a break it will show as either very high or 'infinite'. Repeat with the neutral and the earth.

1 A socket tester is useful as an aid to establish the nature of a fault on a ring final circuit **A**.

2 'Live' testing wth covers removed should always be avoided. The law states that this must only be undertaken with specialist equipment and under special circumstances **B**.

A

B

Make Sure There Are No Short Circuits

A break in a ring will not necessarily stop anything working because current can 'go the other way', but that can lead to part of the circuit carrying too much current and becoming overheated, which in turn can lead to melted insulation and a short circuit. Shorts can also occur where a cable has suffered mechanical damage. To test for short circuits, measure the resistance across live to neutral **A**, live to earth **B** and neutral to earth **C**, they should all be infinite.

You will not necessarily find a partial short circuit with a multimeter. Be aware that a more comprehensive tester may be required.

If everything looks OK, then the MCB itself may be faulty. Make sure the cables are safely out of the way and try switching it back on. If it still trips with nothing attached, fit a new one. See above regarding isolation switches and working on Consumer Units.

If you discover a broken circuit or a short you need to find out where the fault lies. If there has been any other building or DIY work going on, check there first. Nails through cables are a classic and if the nail just penetrates the insulation it may not cause an immediate problem until something bumps against it. If that's not the case, then there is nothing for it but a logical progression around the ring. The process is slightly different for breaks and shorts.

Note: The pictures show tests on an electrically isolated consumer unit, isolated on a separate main switch.

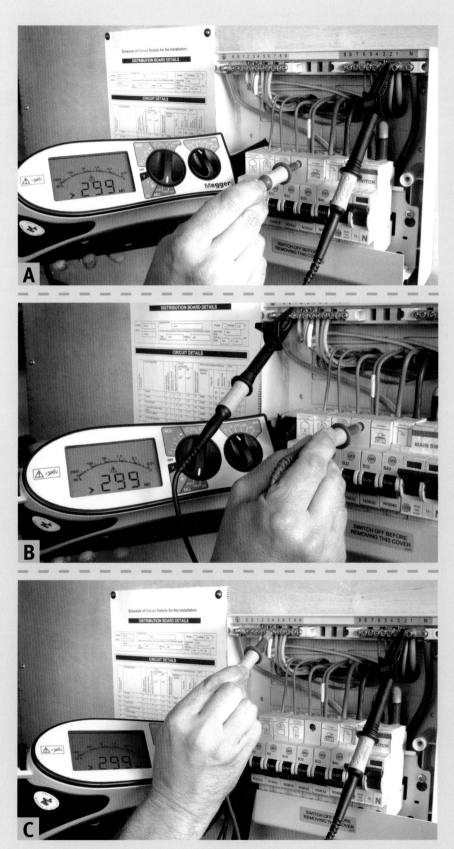

Finding a Break

Isolate the circuit (turn off the supply), twist both the live (line) and the neutral of one leg of the ring final circuit together. Repeat this on the other leg. At a socket the legs of the rings should be disconnected. Using an appropriate tester, test resistance between the live (line) and neutral on each leg, a low resistance should be measured in either direction, that is the twisted conductors of each leg. If an open circuit is recorded on a leg then a break is in the live (line) and neutral is in that direction. Use the same technique to isolate the fault further, by moving to the next socket.

Finding a Short

Leave the power off and make sure all the ends are isolated. Open the first socket and disconnect it. Separate the wires and test for resistance from L-N, L-E and E-N on both cables (or three cables if there is a spur attached). They should all be infinite but one will show low resistance because it is shorted. If that's the cable leading to the CU then you've found the problem. If it isn't, replace all the cables and move on to the next socket. It can take a long time.

Fixing It

Obviously you are going to replace the cable, but you need to find out how it happened to prevent a recurrence. Is the circuit overloaded? Have the correct cables been used? Are the cables run outside of safe zones and subject to damage during other building work? A problem like this can be a timely warning of deeper problems.

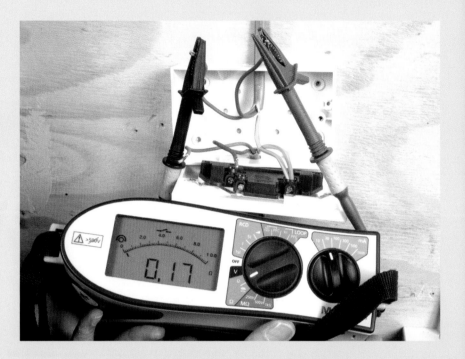

Locating a break in the ring.

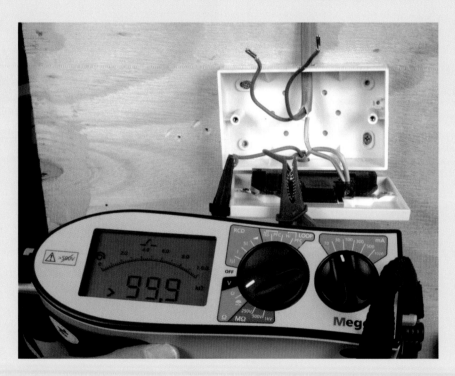

Checking for a short circuit.

No Lights

The most common cause of lighting problems is simply a faulty bulb. As they expire, many cause a brief current surge which is enough to trip the MCB. Usually it happens when you switch a light on so you know which one caused it, but occasionally they die when you're not looking, in which case you just need to turn them all off, reset the MCB and try each in turn. If the MCB won't reset then there is a more serious problem.

Troubleshooting the radial circuit is basically the same as a ring main but with the added complication of ceiling roses (or junction boxes) and switches. Check light fittings carefully first. The heat from a bulb can cause cracks and burnt wires leading to short circuits, especially if powerful bulbs have been used in small fittings. To check a switch, and the cabling to it, disconnect it from the ceiling rose (or junction box) and measure the resistance across the two wires, it should be high or infinite when the switch is off and low or zero when it's on.

No Power At All

Some faults will trip the RCD protecting the whole house supply rather than the MCB for a particular circuit. If it won't reset, turn off all the MCBs and reset the RCD. Assuming it resets OK, reset the MCBs one at a time until one of them trips the RCD. Now you know which circuit is to blame, you can concentrate your efforts on that one. If it won't reset with all the MCBs off, then you probably have a faulty RCD. They are replaceable but if the Consumer Unit is old it might be a good time to replace the whole thing. If the RCD and all your MCBs are OK, the problem is outside your control. Either you have an area fault (ask the neighbours) or the company fuse protecting your house has blown. In either case contact your supplier.

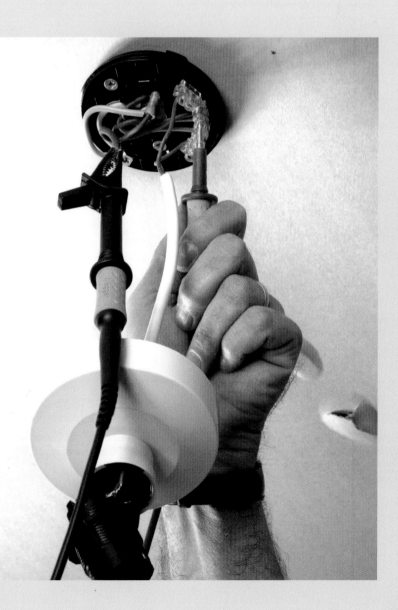

TEST
Measuring the resistance in a switch circuit.

Spurs and Bridges

There is nothing wrong with having some spurs attached to a ring final circuit, but sometimes (especially if the house has been extended or extensively modified) there can be more spurs than ring, and some spurs may have been run from spurs rather than from the ring. The normal 2.5mm² cable used for ring final circuits is suitable for currents up to 20A. Running several 13A sockets on a spur is asking for trouble, so it is worth finding out what is what.

A bridge is where two or more spurs from different parts of the ring have been connected together. This may give the illusion of added safety but can cause localized overloading. In order to test these you need a very accurate resistance meter capable of measuring low resistances with an accuracy of 0.01Ohms. The usual hobbyist multimeter isn't good enough. As these professional meters are expensive it is probably better to get a professional in to test the system for you and give you a report. If you have a suitable meter, here's how to do it:

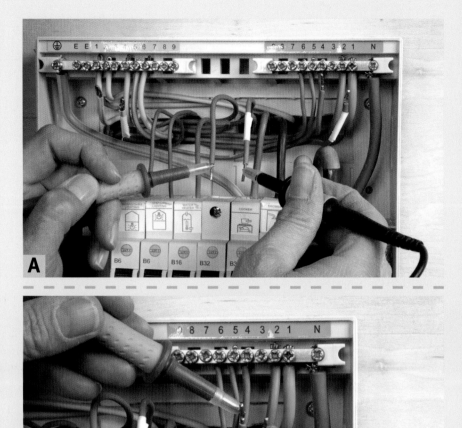

A

B

1 Isolate the supply to the consumer unit. Prove that the consumer unit is dead. Disconnect both ends of the ring final circuit. Test between live and live (also known as line and line) with a low reading ohmeter. Call this r1 **A**.

2 Now test between neutral and neutral, call this rn. The results of these two tests should be the same: that is r1=rn **B**.

3 Now join the line of one leg of the ring final circuit to the neutral of the other leg. Join the remaining line and neutral together. If a test is made at each socket between line and neutral (with a low reading ohmeter), then the values recorded should be the same **C**.

Variations around the ring indicate there is a bridge somewhere, which needs more investigation.

If most sockets show the same resistance, with a few being higher, these are on a spur.

C

The Law and You

Government statistics show that 17 per cent of all fires in domestic premises in England and Wales are caused by electrical faults. In 2004 these resulted in 11 deaths and 1,052 injuries. In addition, 2,809 people were electrocuted, 21 of them fatally. It's not surprising then that the government regulates electrical installation work. Until recently householders were free to carry out whatever work they liked on their own property. They might have had trouble selling the place with poor wiring, and might even have faced negligence charges if their work killed or injured someone, but no one could stop them doing it. The Building (Scotland) Regulations 2004 and the Building Regulations (England and Wales) amended 2005 do not prohibit DIY electrical work, but they do insist that all electrical work is done properly. At the time of writing, Northern Ireland has no such regulations covering electrical work, but for your own safety and to avoid difficulties when selling the property you should make sure your electrical work is done properly.

The following is NOT a complete guide to the regulations, just a simple guide to the parts most likely to affect householders. In case of any queries please contact your local authority's building control department (England and Wales) or Verifier (Scotland).

Scotland

Most building work in Scotland requires a Building Warrant, which the owner should apply for before starting work. If you apply after starting work, or when the job is finished, you will be surcharged. Warrants must be obtained for major electrical work, but there are many jobs you can do without a warrant. Until 2004, warrants could be signed off by anyone with few checks and no requirement that the person certifying the work was qualified to do so. Since 2004 the regulations have been tightened up. If a warrant is issued, it must be signed off either by a qualified tradesperson or a Verifier appointed by the local authority.

In general you do not need a warrant for any wiring or rewiring work in a detached bungalow or two storey house, though you do if the house is higher. You do need a warrant to work in a flat as this might impact on the flats above and below. Semi-detached and terraced houses are more complex. You will need a warrant to do any work that adversely affects a shared wall, such as fitting recessed sockets or chasing a cable run.

If the work does need a warrant, contact your local council for details of how to apply and their scale of charges. Further information is available at: http://www.scotland.gov.uk/Topics/Planning and: http://www.sbsa.gov.uk/

STANDARD
Look for a British
Standard number.

England and Wales

The amended Building Regulations for England and Wales came into force on January 1, 2005. Two sections are of particular interest to DIY electricians, Part P (Electrical safety) and Part L (Conservation of Fuel and Power).

Part P

Part P of the Building Regulations demands that electrical work is carried out by 'a competent person'. Professional electricians should be certified as 'competent' and can issue certificates guaranteeing that their work meets the required standard. Householders can carry out certain simple tasks without certification (see below) but before tackling anything else, you must notify the Building Control department of your local authority.

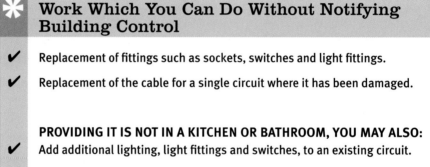

✳ Work Which You Can Do Without Notifying Building Control

✔ Replacement of fittings such as sockets, switches and light fittings.

✔ Replacement of the cable for a single circuit where it has been damaged.

PROVIDING IT IS NOT IN A KITCHEN OR BATHROOM, YOU MAY ALSO:

✔ Add additional lighting, light fittings and switches, to an existing circuit.

✔ Add additional sockets and fused spurs to an existing ring or radial main.

✔ Install additional earth bonding.

EVEN THOUGH THE WORK IS UNLIKELY TO BE INSPECTED, THE LAW INSISTS THAT:

✔ You use suitable cable and fittings for the purpose.

✔ The existing protection measures (earths, fuses and circuit breakers) are unaffected and suitable for protecting the new circuit.

✔ All work complies with all other appropriate regulations.

✳ Work Which Must Be Referred to Building Control

✔ All new wiring

✔ Modifications to wiring in kitchens, bathrooms or shower rooms.

✔ Installation or modification of electric underfloor or ceiling heating.

✔ Garden lighting or power installation.

✔ Other specialist installations such as solar panels and combined heat and power systems.

✔ If in doubt, check with the local Building Control.

How to Apply and What it Costs

As Building Control is run by local authorities, the procedures and costs vary. You can usually download an application form and scale of charges from your authority's website, alternatively write or phone and request a copy. The charges are set locally but regulated by the Building (Local Authority Charges) Regulations 1998. These regulations prevent councils from making a profit from their inspection duties, but do expect them to cover most of the costs of the service.

There is likely to be a sliding scale based on the estimated costs of the work, with a minimum of around £50. This might seem a lot if you are just installing a single new socket in your kitchen but is reasonable when you consider that an inspector will have to visit.

If you are anticipating making many changes it would pay you to plan them all and submit a single notification. Unless you are totally re-wiring the house, the minimum charge will probably still apply and there is no time limit on completion. Providing each stage of the work leaves the whole installation safe, then completing the whole job can drag on for years.

Part L

Part L is concerned with energy efficiency and applies to new buildings and renovation projects, so unless you are self-building, restoring a ruin or converting a barn you probably don't need to worry. If you are doing any of those then you will need to consult your Building Control department on Part L (and many, many other things).

One thing which might concern DIY electricians is the provision of low energy light fittings. Since 2005 it has been compulsory for builders to fit a number of these in each new house or flat, the number varies depending on the size of the property (see table).

Total number of rooms	LE light fittings required
1 to 3	1
4 to 6	2
7 to 9	3
10 to 12	4

Some of these fittings are larger than normal and use larger lamps, so some householders might be tempted to replace these fittings so they can fit conventional lampshades. Some low energy fittings are unsuitable for use with dimmers, which is another reason you might change them.

There is no reason why you shouldn't but be warned, when you come to sell the house you might have a problem with the Home Information Pack surveyor if your fittings are not Part L compliant. If you do change some low energy fittings for conventional ones, refit the low energy versions somewhere else in the house, keeping the overall number compliant with the regulations.

The minimum number is actually very low, and you can have more than one in a single room. Note, if you build an extension you could push the room count into the next category.

For more information about building regulations in England and Wales visit: www.planningportal.gov

Suppliers

www.alertelectrical.com
Very useful website with product details and installation tips.

www.Bdc.co.uk
Distributor of electrical appliances.

www.bewdirect.co.uk
Electrical wholesalers

www.diy.com
Website for B&Q online.

www.esources.co.uk
Wholesale electrical suppliers

www.Maplin.co.uk
Electronics specialists

www.gil-lec.co.uk
Provides electrical accessories

www.homebase.co.uk
Homebase.

www.qvsdirect.com
Provides cable and wiring accessories.

www.screwfix.com
Good for refurbishment jobs.

www.wickes.co.uk
DIY and builders' merchant. Good for the basics, but not as design-led as B&Q or Homebase.

Appliances

www.appliancesdirect.co.uk
Formerly Trade Appliances. Provides electrical equipment and fittings for kitchens.

www.comet.co.uk
Provider of all things electrical for the home and office, including things for the kitchen, computing, entertainment and gadgets.

www.cmsgardens.co.uk
Provides lawn mowers and power tools for the garden.

www.currys.co.uk
Electrical equipment for the home, including dishwashers, TVs, computer, audio and games consoles.

www.lightingstyles.co.uk
Provides lighting for inside and outside the home.

www.low-energy-lighting.com
Supplies low energy lighting units to the public and trade.

www.pcworld.com
Provides computer equipment.

www.richersounds.com
Provider of hi-fi and home cinema equipment.

www.tradingpost-appliances.co.uk
Provides appliances for the kitchen.

Please Note

The publisher is not responsible for the content of external internet sites.

About the Author

Phil Thane has had many jobs, including design technology teacher, software support manager and writer. All have been technical and all have involved explaining technical matters in as clear and concise a manner as possible.

A keen DIYer for many years, he rewired his first house in 1977. These days he lives in North Wales and is gradually creating a terraced garden on a neglected hillside.

Acknowledgements

The author and publisher wish to thank the following for their help in compiling this book:

Ian Archer, Anthony Bailey, Ian Barton, Mike Bartter, the Electrical Safety Council, Dr Dave Kennaird, Peter Upperton, the National Inspection Council for Electrical Installation Contracting (NICEIC), Lydia Morris at Screwfix, Dominique Page, Gill Parris, Hedda Roennevig, Paul Wagland.

Also: Lightyears A/S Balticagade 15, 8000 Aarhus C, Denmark; www.lightyears.dk; and i-Stockphoto.com.

About the Consultant

Dave Kennaird left school at 18 and worked in the electro-mechanical industry designing, building and installing industrial equipment. He then worked for several years for British Telecom before studying for a Mechanical Engineering degree and subsequently a PhD. Dave worked for a number of years as a senior research fellow at the University of Brighton, as well as undertaking consultancy work on a diverse range of engineering projects. Dave now works as a lecturer at Sussex Downs College in Eastbourne.

Glossary

(AC) Alternating Current
Standard mains current in the UK, the voltage on the Live connection alternates from −330V to +330V 50 times every second.

Amp
The unit used to measure current.

Atom
The smallest particle of an element such as copper.

AVO (Amps, Volts, Ohms)
see Multimeter.

Bonding
(see Equipotential Bonding).

Cable
Two or more wires in a single sleeve make a cable.

Conductor
Any material capable of carrying an electric current.

(CU) Consumer Unit
The box next to your meter that houses an RCD and MCBs. See fusebox.

Ceiling Rose
Specialized form of junction box used to hang lights from ceilings.

Current
A measure of the amount of electricity passing through a circuit.

(DC) Direct Current
Current at a steady voltage, unlike AC. Supplied by batteries and most low voltage power supplies for portable equipment.

De-rating
Reducing the allowable load on a device (usually a switch) when used with certain other devices (such as a fluorescent lamp).

DPDT Double Pole, Double Throw
(see below).

Double Pole
A form of switch, resembling two switches operated by a single lever. Used to isolate both L and N connections.

Double Throw
A switch that diverts current from one circuit to another. Commonly called 'two way'.

Earth, Earthing
Safety system used in circuits, literally connected to the Earth to provide a safe path for current leaking from faulty wiring or devices. Vital it is connected correctly.

Electric Current
The movement of electrons through a conductor.

Electrons
Atomic particles carrying a negative electrical charge. Normally, electrons 'belong' to a particular atom but in a conductor they are able to move from one atom to the next, pushed along by the potential difference (voltage) across the conductor.

Equipotential Bonding
Related to Earthing, a method of ensuring there is no potential difference between neighbouring metallic objects. Vital in bathrooms and kitchens.

Flex
Abbreviation of Flexible Cable. Used to connect portable devices.

Fusebox
Early form of CU, using re-wirable fuses instead of MCBs and a simple switch instead of an RCD.

Intermediate (switch)
A form of switch used in 3-way lighting systems.

Isolation (switch)
A double pole switch used to disconnect both L and N connections to a device.

Junction Box
Plastic box, usually round, containing terminals (usually four) for connecting cables.

K
Prefix meaning 1,000. A kilowatt is 1,000 Watts.

Live
In an AC circuit a connection with variable voltage (−330 to +330V in the UK).

Miniature Circuit Breaker
An MCB is a device that automatically switches off the supply when it detects an overload.

Multimeter
Multi-purpose electrical meter capable of measuring voltage, current and resistance.

Negative and Positive

Mathematical convention. Positive numbers are bigger than zero, negative numbers are less than zero. In DC circuits + and − are used to indicate higher and lower voltages and conventional current flows from + to −. It can equally flow from + to 0 or 0 to − or any higher number to any lower one. Electrons carry a negative charge, so their direction of movement is opposite to the 'conventional' flow, but this is only of interest to physicists and electronics engineers! In AC circuits, Neutral is nominally 0V and Live varies from + to − about 50–60 times a second.

Neutral

Connection nominally at 0V in an AC circuit.

Ohm

Unit used to measure resistance. Only relevant if you are using a multimeter to test a circuit.

Potential (Electrical) or (Difference)

Commonly 'voltage', a measure of the difference in electrical charge between two points.

Radial Final Circuit

Normal method of connecting lighting in the UK. Also used for single heavy current devices such as cookers, showers etc.

(RCD) Residual Current Device

An important safety device that compares the current on the Live and Neutral lines and cuts the supply if it detects a difference.

Resistance

A measure (in Ohms) of how easily current flows through a conductor.

Ring Final Circuit (Ring Main)

Normal method of connecting sockets and power outlets in UK homes.

Rose

See Ceiling Rose.

Spur

A branch off a ring or radial circuit to connect a light or power outlet.

Two Gang

Two sockets, switches etc in a single mounting. Three and four gang also used.

Two Way

See Double Throw.

Volt

Unit used to measure potential difference.

Watt

Unit used to measure power. In electrical terms current (Amps) x potential difference (Volts) = Power (Watts).

Wire

A single connector, may be single core (solid) or multi-stranded (flexible).

Picture Credits

All photography by Anthony Bailey/GMC Publications,
with the following exceptions:

Aerolight (www.aerolight.se): 8, 55L

Ballingslov (www.ballingslov.se): 44, 89S, 89R, 98

Bsweden (www.bsweden.com): 6B, 54

The Electrical Safety Council (www.esc.org.uk): 49

Iform (www.iform.net): 5, 67

iStockphoto.com: Jan Neville: 9; Alexandr Tovstenko: 10; Brian Raisbeck:
21; George Peters: 30; Polarica: 31; Daniel Loiselle: 55; Serdar Yagci: 55;
Shelly Perry: 71; Marcin Stalmach: 85; Ugur Bariskan: 87; Sean Boggs: 91;
Konstantin Sutyagin: 93; Andres Balcazar: 95; Forest Woodward: 96; Yury
Zaporozhchenko: 106; Mark Noac: 107; Andrey Prokhorov: 110; Markus
Guhl: 111; Blackred: 111; Pawel Talajkowski: 111.

Lightyears (www.lightyears.dk): 1TL, 1BR, 11L, 11S, 11R, 20, 61, 89L, 111L,
111S

Marbodal (wwwmarbodal.se): 2, 3

The Pylon Appreciation Society (Flash Wilson Briston) www.gorge.org: 14

Screwfix (www.screwfix.com): 28, 29, 30, 31, 33, 34, 36, 37, 38, 39, 40, 41,
42, 43, 45, 46, 47, 86, 92, 93TR, 93BL, 97, 102

Vola (www.vola.com): 4, 7T, 88, 90, 92

Wavin (www.hepstore.co.uk): 84

T=top, M=middle, B=bottom, L=left, R=right, S=second, TH=third

Useful Websites

The National Inspection Council for Electrical Installation Contracting (NICEIC), the body that regulates the professionals, has useful information for amateurs too and publishes detailed guides. Essential if you are contemplating a major re-wiring job.
www.niceic.org.uk

The Electrical Safety Council publishes guidance notes.
www.electricalsafetycouncil.org.uk

Information about testing can be found at:
www.electrical-testing-safety.co.uk

Information for children on how electricity works and how to use it safely:
www.switchedonkids.org.uk

For a guide to ensuring electrical safety in your home, see:
www.direct.gov.uk

For information on electrical safety regulations log on to:
www/planningportal.gov.uk

For DIY articles on home wiring see:
www.diydoctor.org.uk
www.diyfixit.co.uk
www.doityourself.com

For information on training see:
www.niceic.org.uk/niceictraining

For information on electrical safety equipment in the garden see:
www.letsgogardening.co.uk

For information on buying recommended energy saving products, see:
www.energysavingtrust.org.uk

For information on the Electrical Contractors Association log on to:
www.eca.co.uk

For information on the Instituition of Engineering and Technology, see:
www.theiet.org

Note
The publisher is not responsible for the content of external internet sites.

Index

To request a full catalogue of GMC titles, please contact:
GMC Publications
Castle Place
166 High Street
Lewes
East Sussex
BN7 1XU
United Kingdom

Tel: 01273 488005
Fax: 01273 402866

Website: www.thegmcgroup.com